CALIFORNIA NATURAL HISTORY GUIDES

GEOLOGY OF THE
SAN FRANCISCO
BAY REGION

Dear Morton,

Enjoy reading about
our interesting geology
and exploring our
beautiful parks. I enjoyed
visiting with you at the
City Commons Club.

John Karachewski
5-14-2010

California Natural History Guides

Phyllis M. Faber and Bruce M. Pavlik, General Editors

GEOLOGY

of the SAN FRANCISCO BAY REGION

Doris Sloan

With photographs by John Karachewski

UNIVERSITY OF CALIFORNIA PRESS

Berkeley   Los Angeles   London

Dedicated to the late Clyde Wahrhaftig, mentor and friend, who taught me how to read the landscape.

University of California Press, one of the most distinguished university presses in the United States, enriches lives around the world by advancing scholarship in the humanities, social sciences, and natural sciences. Its activities are supported by the UC Press Foundation and by philanthropic contributions from individuals and institutions. For more information, visit www.ucpress.edu.

California Natural History Guide Series No. 79

University of California Press
Berkeley and Los Angeles, California

University of California Press, Ltd.
London, England

© 2006 by the Regents of the University of California

Library of Congress Cataloging-in-Publication Data

Sloan, Doris, 1930–.
   The geology of the San Francisco Bay region / Doris Sloan ; photography by John Karachewski.
      p. cm. — (California natural history guides ; 79)
   Includes bibliographical references and index.
   ISBN-13, 978-0-520-23629-5 (cloth, alk. paper), ISBN-10, 0-520-23629-7 (cloth, alk. paper)
   ISBN-13, 978-0-520-24126-8 (pbk., alk. paper), ISBN-10, 0-520-24126-6 (pbk., alk. paper)
   1. Geology—California—San Francisco Bay Area. I. Title. II. Series.
   QE90.S12S58 2006
   557.94'61--dc22

2004014703

Manufactured in China
16 15 14 13 12 11 10 09 08
10 9 8 7 6 5 4 3 2

The paper used in this publication meets the minimum requirements of ANSI/NISO Z39.48–1992 (R 1997) (*Permanence of Paper*). ♾

Cover: View south across Golden Gate toward the Presidio of San Francisco. Ridge of Franciscan chert in foreground. Photograph by John Karachewski.

The publisher gratefully acknowledges the generous
contributions to this book provided by

the Gordon and Betty Moore Fund
in Environmental Studies
and
the General Endowment Fund of the
University of California Press Foundation.

CONTENTS

ACKNOWLEDGMENTS

The inspiration for this book came primarily from my students in classes on Bay Area and California geology, who demanded clear and simple explanations for a geology that is neither. It is a collaborative effort in many ways. The fine photographs by John Karachewski are integral to the book. The geology would not be clear without them, and I am grateful for his work, as well as for the pleasure it was to work with him on this project. I thank my colleagues in the Department of Earth and Planetary Science (formerly Geology and Geophysics) at the University of California at Berkeley for introducing me to the joys of playing geologic detective. I am particularly indebted to Garniss Curtis, whose willingness to share his broad knowledge of the geologic landscape has been an inspiration to me for more than three decades.

This story of Bay Area geology could never have been written without the work of the many colleagues who have explored its rocks and landscapes and who have helped me with information and criticism as I struggled to write clearly about our geology. My deepest gratitude goes to both faculty and students at the University of California at Berkeley and in the earth sciences departments at the other Bay Area academic institutions, to colleagues at the U.S. Geological Survey in Menlo Park and at the California Geological Survey in Sacramento, and to those in independent practice. Without their significant contributions, this book would not have been possible. Special thanks to Russell W. Graymer at the U.S. Geological Survey in Menlo Park for giving unstintingly of his time and knowledge, especially with the geologic maps. Mel Erskine also deserves special thanks, not only for reading the manuscript, much of it more than once, but also for the many hours he spent discussing the best ways to simplify a complex story.

I want to thank the many friends who were my companions and chauffeurs on field explorations to the far corners of the Bay Area, including Ed and Loralee Hiramoto, Diana King, Ron and Elena Krause, Dorothy Lindheim, Vickie Robinson, and Sharie Shute. My gratitude also to those who read the manuscript. John Wakabayashi reviewed an early draft and made very helpful comments; his thorough review improved the manuscript a great deal. David W. Andersen, Jane Brandes, Garniss Curtis, Phyllis Faber, Russ Graymer, David Howell, James Ingle, Diana King, Louise London, Vickie Robinson, Douglas and Norah Rogers, Andrei Sarna-Wojcicki, Nicki and Tom Spillane, and Stephen Vonder Haar read all or part of later versions. They have given me invaluable feedback and have prompted many corrections. All remaining errors are mine.

I was helped immeasurably by staff of the Department of Earth and Planetary Science, the Department of Geography, and the Earth Sciences Library, all at the University of California at Berkeley. Margaret Gennaro took the fine photographs of rocks and minerals; Charley Paffenberger and Dino Bellugi printed maps; and Don Bain sharpened aerial photographs.

Thanks also to Doris Kretschmer, Scott Norton, and Kate Hoffman of the University of California Press for their help in all phases of preparing this book. Without their patience, advice, and encouragement the book would never have been finished. And my heartfelt thanks to my family for their loving support through the long process of creating this book.

INTRODUCTION

Geology is a lens through which we can see the world around us in a new dimension. Knowing the name of a wildflower makes it our friend. Knowing that the bulbous black rock at Point Bonita was formed in the deep sea and has traveled thousands of miles across the Pacific, makes it part of us. So, too, understanding why Red Rock is red, why Loma Prieta is high, why a bit of ocean floor lies on top of Mount Diablo, enriches our connection to the earth. This book is for the San Francisco Bay Area resident or visitor who wants to explore the geologic landscape of the Bay Area, who is curious about its shapes, colors, and rocky foundation. "San Francisco Bay Region" is the more formal term for the nine counties that touch San Francisco Bay. Informally, most residents of the region refer to "the Bay Area." In this book, the former term is used for the title, the latter throughout the text.

I began my study of the geology of California and the Bay Area 30 years ago and quickly came to appreciate their considerable geologic complexity. At that time, plate tectonics was still relatively new and not universally accepted. In the past three decades, both understanding and complexity have grown. Exciting new interpretations continue to be put forth. That is one of the marvelous things about geology—it is not static and never boring. It changes continually as new eyes look at the rocks, faults, and hills, and new technologies increase our understanding of what we see. If you first read about Bay Area geology 10 or 20 years ago, it is time to take a new look at the story of the processes that created our present landscape, and the Bay Area's geologic history.

Keep in mind that geology is a dynamic science. Ongoing research leads to new concepts and interpretations. This process is continual, and undoubtedly some of the material in this book will become out-of-date. I have attempted, of course, to be as accurate and current as possible, but my colleagues will certainly discover many new and exciting aspects of Bay Area geology in the coming years. Remember also that geologists often differ about interpretations of what they see. One of our favorite pastimes is to stand at a rock outcrop and argue about what it means. Do not treat anything in this book as gospel; consider it a work in progress, like our understanding of the complex and fascinating geology of the Bay Area.

What This Book Covers

This book considers the geologic underpinnings of the Bay Area landscape and the processes that have shaped it. Most of us become geologists because we love the out-of-doors and equally love a good detective story. The mystery of how our landscape formed compels us to poke around the far corners of the Bay Area, exercising both body and mind, in order to observe the many processes that created the landscape.

Rocks are the chief clues to understanding the geologic history of the Bay Area. We whack them with geologic hammers, peer closely at them through a hand lens, take samples to determine how old they are, and make thin slices to study their chemistry under the microscope. Through the rocks we then put together the story of the history and landscape. In this book I share with you what hundreds of geologic detectives have learned over more than a century of study of Bay Area rocks.

Chapters 1, 2, and 3 provide some background information on geologic processes, time, and rocks that is important to an understanding of the Bay Area landscape and geologic history. The rest of the book covers the region geographically, counterclockwise, starting in Marin County. We start there because the rocks in Marin provide a good introduction to the story of the plate tectonic beginnings of the Bay Area. Each chapter starts with the present landscape, then discusses the tectonic structure of the area, its rocks, and history.

A word about what this book is not. First, it is not a guide to particular sites, although I give many examples, especially from parks and public open space, to illustrate the geology. Several geologic field guides intended for the general reader are available, and local geological associations publish field guides regularly, although most are for a technical audience. They are listed under Further Reading. Second, this book is not the story of the development of the Bay Area landscape over time, although I have tried to give the reader a general sense of this story in each chapter. That full reconstruction awaits another book.

Politically, the entity we call the Bay Area encompasses the nine counties that touch the Bay: Alameda, Contra Costa, Marin, Napa, San Francisco, San Mateo, Santa Clara, Solano, and Sonoma. Geologically, the Bay Area, of course, is defined not by political borders but rather by watersheds, faults, and rock types. This book covers the nine counties and is divided into geographic sections that are partly bounded by the region's natural features. To a surprising extent the geology differs in each section, and the divisions of the book would be similar even if political boundaries were ignored. The western margin of the Bay Area is the Pacific Ocean; the eastern is the Delta. Northern and southern boundaries of the Bay Area are arbitrarily taken at county lines, which is expedient but not entirely satisfactory. Now and then we will cross a county line to look at interesting geology just beyond it. And when we talk about San Francisco Bay, for example, we really have to think as far east as the Sierra Nevada, the source of most of the water that flows through the Bay Area.

This book is intended as a beginning for your exploration of Bay Area geology. Of necessity it is brief and omits a great deal. I hope it leaves you thirsting for more information. Among the resources for further exploration of Bay Area geology given in Further Reading, I want to point out two books that are classics of Bay Area geology. One is the *Geologic Guidebook of the San Francisco Bay Counties,* published by the California Division of Mines (now California Geological Survey) as Bulletin 154 in 1951. It is valuable not just geologically but also historically. Although some of the geologic interpretation is now out-of-date (it was written before the development of plate tectonics), Bulletin 154 is mostly written in nontechnical language and is full of interesting and unchanging information on landscape, fossils, and minerals. It is out of print, but if you see a copy in a secondhand bookstore, snatch it up for a great deal of reading enjoyment and some interesting field excursions. The other book is *A Streetcar to Subduction,* by the late Clyde Wahrhaftig, one of the Bay Area's outstanding field geologists. Written in 1984, it presents plate tectonic interpretations of Bay Area geology and field trips to key locales. Several of the field sites are no longer available, but the book will give you a valuable understanding of many fine sites. Both of these books are excellent additions to a library of Bay Area geology.

Where to Begin

You do not need much equipment to enjoy geology, but a 10-power hand lens or other strong magnifying glass is a wonderful thing to have along on your explorations. It allows you to see the individual grains of beach sand in all their variety and crystals on a rock that sparkle in the sun. You may enjoy taking along a geologic map. The U.S. Geological Survey in Menlo Park has issued a fine series of geologic maps that cover most of the Bay Area (see Further Reading). For most exploration of Bay Area geology you will not need that emblematic geologic tool, a rock hammer. Its use is not permitted in parks and preserves, and unless one is very conscientious, whatever rock is hammered is also defaced.

Understanding the geology of any area is an iterative experience, especially in areas as geologically rich as ours. Some geologic processes are direct and easy to understand, such as the action of waves along the coast. Others are complex and will require repeated exploration and reading. Your efforts will be rewarded by a deeper appreciation of the magnificent geologic landscapes of the Bay Area.

A Note about Geologic Maps and Rock Names

The colors on geologic maps, such as map 2 and the maps that accompany each regional chapter, indicate the type of rocks exposed at the surface if you remove the cultural features. The colors convey information about the age and composition of the rocks. Relative age is given by their name or shown by the order in which the rocks are listed in the legend: oldest at the bottom, youngest at the top. Rocks of different composition are shown by different colors. Geologists have developed a general color style that is followed as much as possible in this book: granitic rocks are pink; volcanic rocks are orange; very young sedimentary rocks are light yellowish colors, and older ones are darker; the state rock, serpentinite, and its associated rocks are purple; and so on.

On most of the geologic maps in this book, the different rock units are listed by age and composition, for example, Tertiary volcanic rocks. On more detailed maps, such as Map 8 of Point Reyes, they are listed by name, for example, Millerton Formation. Rocks of a similar age and geologic origin are grouped as a "formation" and given a name that is usually based on a nearby geographic feature. For example, the Monterey Formation is a sedimentary rock that is common on the Monterey Peninsula. A group of rocks that differ in composition but are related in origin is referred to as a "complex," for example, the Franciscan Complex, which consists of volcanic, sedimentary, and metamorphic rocks formed and brought together by plate tectonic processes. On the geologic maps, the most common Franciscan rocks, pillow basalt, graywacke, shale, and chert, are referred to as "Franciscan Complex coherent rocks" to distinguish them from the ground-up rocks of the Franciscan mélange (see chapter 3).

SOME OF THE MOST INTERESTING geology in the world is found in our backyard. Here you can see almost every type of geologic process in action. The Bay Area is a colorful geologic mosaic, its tiles composed of a great variety of rocks, its patterns arranged and then rearranged by active faulting. Soften the edges of the tiles through the pounding of the ocean waves or the slump of a landslide, put a magnificent bay into the center of the mosaic, and you have a world-class geologic setting.

The Bay Area (map 1) holds many geologic surprises. The familiar hills and valleys, which to our "rock of the ages" mindset seem to have been here forever, are mere geologic infants. Get into the family time machine and visit the Bay Area of a million years ago and you see a far different landscape. Take your time machine a million years into the future, and it is likely you will not recognize your homeland. The mountains that look so solid, so unchanging, are rising and eroding rapidly. Today's landscape is just 1 or 2 million years old—very old in human terms but young to a geologist. The present landscape is temporary, here today but gone from the geologic tomorrow.

And San Francisco Bay—how deep and permanent its waters look! But if you were 12 feet tall, you could walk across much of it with your head above water. Nor has it been here long. Take your time machine back 20,000 years to the time of the last glaciation and you see only a wide grassy valley, not a bay at all.

The variety of rock types in the Bay Area is also a surprise (map 2). In comparison with, say, the layer-cake geology of the Grand Canyon, we have a great many more kinds of rocks, and some are highly unusual. This is true of West Coast rocks from Mexico to Alaska. These rocks look firmly in place, yet many have traveled great distances in time and space to get here—millions of years and thousands of miles. Some, like our blue-green California state rock, serpentinite, originally came from far below the earth's surface. Millions of years of movement along the Bay Area's many faults have rearranged the rocks into a geologic complexity that defies order and reason.

Bay Area geology is so uncommonly interesting and complex because we live in one of the most exciting geologic settings on earth, a place where two of the tectonic plates that make up the outer shell of the earth meet. Dynamic geologic action takes place where these plates move against each other, causing earthquakes and volcanoes, building mountain ranges, and concentrating mineral resources. The concept of plates and their movement, called plate tectonics, is central to understanding Bay Area geology. Almost every aspect of its geology, including topography, rocks, and natural hazards, is affected in some way by plate tectonics. Chapter 2 considers the process in more detail.

Time: Human and Geologic

To understand Bay Area geology, one needs to be aware of the processes that are shaping its landscape today and also what has happened in the past. Processes acting today at the *human* time scale are not too difficult to comprehend; however, understanding events that happened long ago on a *geologic* time scale requires a mental exercise we do not often get. We are comfortable with decades and centuries but have little feel for a thousand years, let alone a million. Yet to a geologist, a thousand years is an instant of time, and even a million years is not very long.

To get a feel for geologic time, consider this analogy. If you compress the 4.6-billion-year age of the earth into 1 year, a month equals about 380 million years; a day, 12 million years; and each hour, 500,000 years. At this scale, the earth's oldest known rocks (4.1 billion years old) formed in mid-February, the oldest California rocks (1.7 billion years old) in mid-August, and the oldest California fossils (1.3 billion years old) in late September. The Paleozoic Era (when animals with hard shells first appeared) begins in mid-November, the Mesozoic (the age of dinosaurs) in mid-December, and the Cenozoic (the age of mammals) about December 26. The key geologic processes that form the rocks of the Bay Area begin in mid-December, though a few of our rocks may be as old as mid- to late November. Much of the Bay Area landscape begins to form a few hours before the end of the year, and modern humans come on the scene a few minutes before midnight, December 31.

Familiarity with the geologic time scale (fig. 1) will make it easier for you to follow the changes over geologic time. The oldest common rocks in the Bay Area—the Franciscan, Great Valley, and Salinian Complex rocks—were formed or assembled primarily during the Mesozoic Era. Many of the widespread
text continues on page 8

Map 1. Map of the Bay Area.

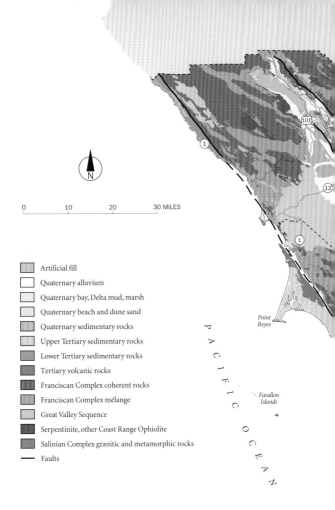

Artificial fill
Quaternary alluvium
Quaternary bay, Delta mud, marsh
Quaternary beach and dune sand
Quaternary sedimentary rocks
Upper Tertiary sedimentary rocks
Lower Tertiary sedimentary rocks
Tertiary volcanic rocks
Franciscan Complex coherent rocks
Franciscan Complex mélange
Great Valley Sequence
Serpentinite, other Coast Range Ophiolite
Salinian Complex granitic and metamorphic rocks
—— Faults

0   10   20   30 MILES

N

PACIFIC OCEAN

Point Reyes

Farallon Islands

Map 2. Geologic map of the Bay Area.

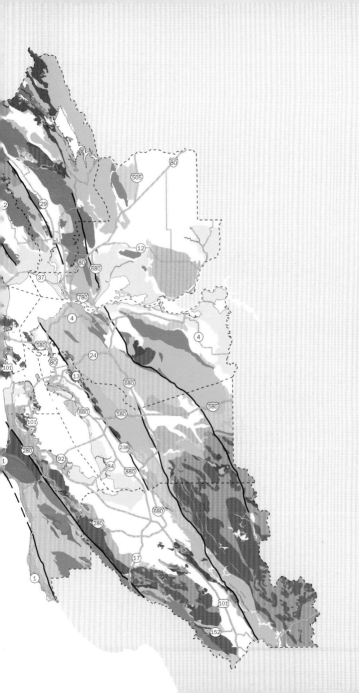

sedimentary and volcanic rocks were formed in the Tertiary Period of the Cenozoic Era. Many of the youngest geologic materials, those of the Quaternary Period, are not yet rock. Sediments accumulating today in lakes and along rivers, the beach sands and cobbles, landslide debris, San Francisco Bay mud, and other surface deposits are the raw materials for the rocks of the future.

MYA	Era	Period		Epoch	Duration (MY)
0.01	Cenozoic	Quaternary		Holocene	0.01 (10,000 years)
1.8				Pleistocene	1.8
5.3		Tertiary	Late	Pliocene	3.5
23.0				Miocene	17.7
33.9			Early	Oligocene	10.9
55.8				Eocene	21.9
65.5				Paleocene	9.7
145.5	Mesozoic	Cretaceous			80.0
199.6		Jurassic			54.1
251.0		Triassic			51.4
299.0	Paleozoic	Permian			48.0
318.1		Pennsylvanian			18.1
359.2		Mississippian			41.1
416.0		Devonian			56.8
443.7		Silurian			27.7
488.3		Ordovician			44.6
542.0		Cambrian			53.7
4,600	Precambrian	Origin of Earth			~ 4,000

Figure 1. The geologic time scale. MY = million years, MYA = million years ago.

Some geologic processes occur quickly, for example, earthquakes, floods, and many landslides. Other processes operate over hundreds to millions of years, such as the uplift and wearing down of hills, the movement of tectonic plates. Even a very slow process can bring great changes, given a long enough time. The Bay Area did not always look like it does today. Warm up your time machine and go back 150 million years. Looking out the window you do not see any land, only the waters of a vast ocean. The shoreline is far to the east, where the Sierra foothills are now or even farther east, and great volcanoes like those in the Cascade Range or Andes are erupting in what today is eastern California. Now travel to about 10 million years ago, and you see land, but not as it is today. A long ridge of hills lies where you expect to see the bay. The East Bay Hills do not exist; Mount Diablo is not there. Next set your time machine to about 1 million years ago, and you can see the East Bay Hills and Mount Diablo beginning to rise. The present landscape is taking shape.

The geologic processes that shaped the landscape may be very different from those that formed the rocks in that landscape. Many Bay Area rocks are many millions of years old, formed when this area was under water. Thus, when you look out over today's landscape, you should ask yourself two questions: How and under what conditions were the rocks formed? and, What processes shaped the landscape? The answers usually involve quite different processes.

It takes practice to get a feel for geologic time. Today's landscape and rocks are the result of processes that operate on both human and geologic time scales. Try to keep these two perspectives in mind as you look at local geology. Remember to ask, What processes are going on today? and What happened long ago?

In order to understand the flow of geologic processes, we need some understanding of the ages of the rocks. Rock units can be dated in relative (older, younger) terms by their relationships to each other and by the fossils in them. In the Grand Canyon, the Colorado River has cut down through many different types of rocks. The upper 4,000 feet consists of horizontal layers that reflect the order in which they were originally formed. Therefore, we can say that the higher layers are younger, relative to the lower ones. Until the beginning of the twentieth century, relative dating was the only method available. Then techniques utilizing the radioactivity of certain elements led to absolute dating, in which an age in years can be assigned to rocks containing those radioactive elements. Thus, we know, for example, that the granitic rocks at Point Reyes were formed 80 to 90 million years ago. Many rock units are formed over a considerable span of time, and for many we do not have precise ages. Because of this, geologists may speak of a rock as formed, for example, in the Miocene (Epoch) or the Paleozoic (Era). So keep the geologic time scale handy for reference as you explore Bay Area geology.

Geologic Processes: Shaping the Landscape

Geographically, the Bay Area is in the Coast Range Province, one of the 11 geomorphic provinces into which California is divided (map 3). Each province has its distinct geologic style and history. The Coast Range Province extends 400 miles along the coast of California, from the Transverse Ranges in the south to the Klamath Mountains in the north. It is bounded by the Central Valley Province on the east and the Pacific Ocean on the west. Unlike the Sierra Nevada, which is one long mountain range, the Coast Ranges are a series of more or less parallel ranges, all running generally northwest to southeast.

What we usually notice first about any landscape is its topography: the pattern of hills and valleys, the streams and lakes, the shape of the coastline. When we look at the Bay Area from space we can see that it is a hilly place (pl. 1). The many hills and wide valleys generally run northwest to southeast, like California itself. They are modest hills compared to the Sierra Nevada, though we call them mountains. The highest are Mount Tamalpais in Marin County at 2,604 feet, Mount Diablo in Contra Costa County at 3,849 feet, Mount St. Helena, where Sonoma, Napa, and Lake Counties meet, at 4,344 feet, and Copernicus Peak in Santa Clara County at 4,373 feet, a little higher than nearby Mount Hamilton's 4,209 feet. The largest lowland, a wide valley that extends almost the length of the Bay Area, is partly filled by San Francisco Bay. We can see the mighty Sacramento–San Joaquin River system flowing into the bay through Carquinez Strait and out to the ocean through the Golden Gate. The Russian River cuts a wide

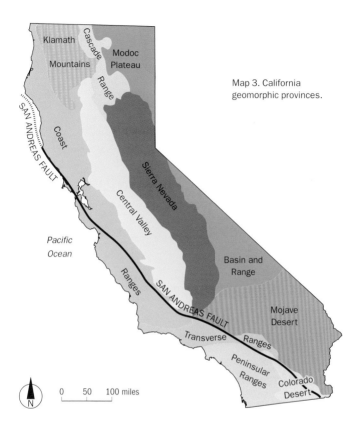

Map 3. California geomorphic provinces.

valley as it flows south into the Bay Area and then makes an abrupt turn westward to the ocean. On the west is a long coastline, steep and rocky in some places, with wide beaches or calm bays at others.

The Bay Area landscape is shaped by a variety of geologic processes. They include tectonic activity, which moves pieces of the crust up, down, and sideways; weathering and erosion, which wear the high places down; deposition of sediment eroded from the high places; and coastal processes where land meets sea. Many of the geologic processes that operated in the past are still going

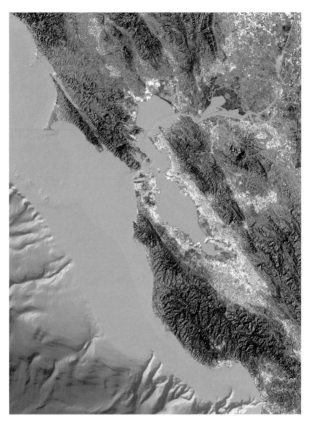

Plate 1. Topography of the Bay Area.

on today. By observing them we can see how they change a landscape. Others are no longer active here, and we have to look in other parts of the world to see how they have contributed to Bay Area geology. For the past several hundred years, humans, too, have modified the landscape by leveling mountains, constructing roads and structures, damming rivers and filling the wetlands at the edge of the bay. All of these processes, natural and human, contribute to the mosaic of Bay Area geology. With our rocks as clues, we can look into the past and see what has come before.

At the Surface: Weathering and Erosion

The natural process of altering and breaking up solid rock into smaller and smaller fragments, either chemically or mechanically, is weathering. Chemical weathering changes the minerals in rock or dissolves them. Mechanical weathering physically breaks up the rock. Living organisms, ranging in size from microbes to tree roots, often are agents of both chemical and mechanical weathering.

One of the first steps in weathering is oxidation of iron-bearing minerals in the rock. The rock becomes more brown as the iron oxidizes (pl. 2). Because most rocks contain some minerals with iron, brown color is not a reliable indicator of their composition. Weathering also builds soils. The characteristics of soils, such as color and nutrients, come in large part from the nature of the underlying rock.

The process of erosion carries the products of weathering from high places to lower elevations, usually through the action of water, and deposits them as sediment in river beds or lakes and ultimately in the ocean. Along the way, the particles may rest on a hillside or alluvial plain, such as the broad, gently sloping area between the hills and San Francisco Bay. Wherever the products of weathering and erosion are deposited, they are initially loose or

Plate 2. Weathered (brown) and fresh (gray) granitic rocks at Montara Mountain, San Mateo County.

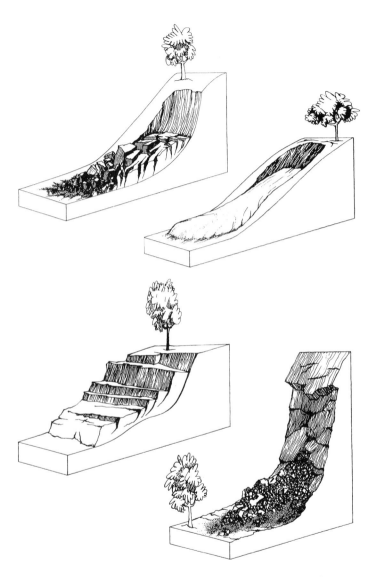

Figure 2. Types of landslides: *upper left*, debris slide; *upper right*, earth-flow; *lower left*, slump; *lower right*, rockfall.

Plate 3. A 1998 landslide on Mission Peak, Alameda County.

unconsolidated. Over time, as they are buried by more deposits, they get compacted or cemented, or both; that is, they lithify and turn into rock. This rock may then be lifted up, weather, and erode, as part of the endless rock cycle.

The hardness and composition of a rock, as well as its physical setting, affect the rate of erosion. Generally, where the rocks are hard and resistant, erosion rates are relatively slow; where the rocks are softer, rates can be rapid. Though much erosion operates over geologic time, landslides are dramatic reminders that geologic change can also take place in a very short time frame and have catastrophic consequences.

Four types of landslides are common agents of erosion in the Bay Area (fig. 2). In a debris slide, the landslide consists of incoherent or broken material; in an earthflow, water-saturated material flows downhill like a very thick fluid; a slump consists of coherent material that rotates as it moves downhill; a rockfall consists of material that primarily falls through the air. Many slides are a combination of these types (pl. 3). Earthflows are common in the Bay Area. They produce a characteristic hummocky ground surface that has been called "melted ice cream topography" (pl. 4). It is especially easy to see when sunlight is at a low angle in the morning or evening. Once you learn to recognize such a surface, you will see it on many Bay Area hillsides and will understand the large role this type of landslide plays in shaping

Plate 4. "Melted ice cream topography" on Rocky Ridge in Las Trampas Regional Wilderness, Contra Costa County.

the Bay Area landscape. Earthflows are a major cause of damage to property, especially in wet winters. The great need for housing in the Bay Area has sometimes led to inadequate consideration of the hazards of construction in areas prone to sliding. Rockfalls are common along the coast and along steep road cuts. Inland, they are less of a hazard than the other types of landslide because they tend to form on steep slopes, where construction is unlikely.

A hazard in young unconsolidated sediments is liquefaction during an earthquake. Where such sediments are saturated with water, shaking can turn them into a slurry that is unable to support the weight of structures such as buildings, bridges, highways, and airport runways. Liquefaction hazard is particularly acute around the margin of San Francisco Bay, where much damage has occurred during large earthquakes.

Where Sea Meets Land

California has over 1,000 miles of coastal shoreline with a great variety of rock types and magnificent scenery. The dramatic views are equaled only by the dynamic geologic processes that shape this landscape. Although the same geologic processes operate at the shore as inland, two other factors are important along the coast: wave action and changes in sea level. Weathering along the coast is rapid where rocks are soft or much fractured, and slow where they are hard and resistant. Because of the wide variety of rocks exposed along the Bay Area coast, erosion rates vary greatly from place to place. But everywhere the pounding of the waves takes its toll on the land.

Erosion by waves takes several forms. The mechanical force of crashing waves fractures the rock into increasingly smaller fragments. The force of the water compresses air in cracks in the rock and breaks it apart to form caves and arches, especially along fractures where the rock is already weak. Erosion of a sea cliff is speeded up as its base is undercut by the waves. Where beaches are wide and waves do not break against the cliff, erosion is slower. Sea stacks, a dramatic element of our coast (pl. 5), are remnants of the most resistant rocks, left behind as the softer rocks were washed away.

Plate 5. Sea stacks in Franciscan rocks, south of Russian River, Sonoma Coast State Beach.

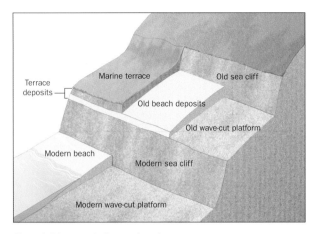

Figure 3. Wave-cut platform and marine terrace.

Plate 6. Wave-cut platform at James V. Fitzgerald Marine Reserve, San Mateo County.

As waves erode the edge of the land, they form a wave-cut platform, a gently sloping surface of rock that extends from the beach out into the ocean (fig. 3). Wave-cut platforms are cut at times when sea level is high and relatively stable for an extended time. Sea level is high today, as it was during the interglacial times about 85,000, 105,000, and 120,000 years ago, and earlier, because less water is present on land in glaciers and ice sheets. Commonly the wave-cut platform is buried by sand or gravel beach deposits, but along some shores in the Bay Area you can see the platform at a low tide. The James V. Fitzgerald Marine Reserve north of Half Moon Bay in San Mateo County has a good example (pl. 6).

Along tectonically active coasts like those of the Bay Area, the wave-cut platform may be uplifted to form a relatively flat area called a marine terrace (fig. 3). Then a new wave-cut platform is carved at sea level. You can see excellent examples of marine terraces along the coast in the Bay Area (see chapters 4, 7, and 10). The terraces commonly are covered by beach deposits and by sediment eroded from uphill. Where uplift occurs repeatedly, several levels of marine terraces, each progressively older and higher, can be recognized landward of the shore. Measurements of coastal uplift at various locations in the Bay Area indicate that the land presently is rising at a rate of about 1 to 3 inches a century. Although this may seem like a very slow rate, even fractions of an inch per year operating over a long time can make major changes in the landscape.

The Sand on (and off) the Beach

Most beach deposits have a local source; thus, if you compare sand or pebbles from different beaches in the Bay Area, you will find a great range in size, color, and composition of the grains, depending on the surrounding rock (pl. 7). Take a magnifying glass or hand lens when you go to the beach, and enjoy the beauty of the different sand grains. Beach deposits are commonly sorted by grain size and density as they are moved back and forth by the waves. Vigorous winter waves can move quite large cobbles and can transport large quantities of sand off the beach, depositing it several hundred yards offshore. Gentle summer waves gradually return the offshore sand to the beach. As a result, beaches change in size and shape seasonally; a sandy beach in winter tends to be narrower and lower than in summer, when sandy beaches are at

Plate 7. Variety of beach sands. *San Mateo County:* A, Pigeon Point; B, Pescadero State Beach; D, Thornton Beach; H, El Granada. *San Francisco County:* E, Ocean Beach. *Marin County:* C, Cronkhite Beach; F, Stinson Beach. *Sonoma County:* G, Bodega Head.

their widest because of seasonal onshore movement of sand. During years of major winter storms, much sand can be removed from a beach. The El Niño storms of 1983 removed many feet of sand from Bay Area beaches (pl. 8). Gradually, over several years, the beach sand is replenished.

Beach sands are also moved along the shore by longshore drift (fig. 4). Where the prevailing current hits the shore at an angle, the current moves sand slowly down the coast. In the Bay Area, sand is moved generally southward. Where waves are refracted around headlands, these currents form sand spits, barrier beaches, and bay-mouth bars. Limantour Spit at Point Reyes and Doran Park at Bodega Head (pl. 9) are fine examples. Some

Plate 8. House at Seadrift left up in the air by El Niño storms of 1983. Note stairs that formerly ended at beach. Seadrift is a sand spit across the San Andreas Fault Zone at Bolinas Lagoon in Marin County.

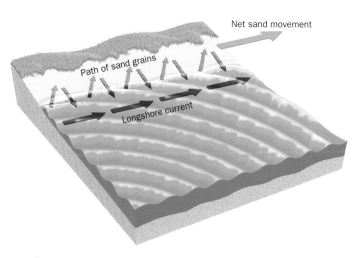

Net sand movement

Path of sand grains

Longshore current

Figure 4. Longshore drift.

Plate 9. Sand spit at Doran Regional Park, looking east to town of Bodega Bay, Sonoma County.

Plate 10. Cusps at Bakers Beach, Golden Gate National Recreation Area, San Francisco.

beaches, such as the Point Reyes beach and Bakers Beach in San Francisco, develop prominent pointed cusps that show how high-energy waves move the sand around (pl. 10). These features can be seen along the Bay Area's long shoreline, a magnificent natural laboratory for exploring coastal geologic processes.

All the surface processes—weathering, erosion, wave action —work together to shape our landscape. The tectonic processes that form and rearrange other elements of the landscape are considered in the next chapter.

PLATE TECTONIC PROCESSES are fundamental to the geology of the Bay Area. Every aspect of our landscape—the hills and valleys, the underlying rocks, the earthquakes that occasionally shake us—is the result of living at an active boundary between the North American and Pacific tectonic plates. Because much of the dynamic geology on our planet takes place at plate margins, we have a front-row seat in the tectonic theater. We live along the "Pacific Ring of Fire," the zone of active earthquakes and volcanoes that circles the Pacific Ocean.

"Tectonics" refers to movements of the earth's lithosphere, which consists of the solid crust and the uppermost solid part of the underlying mantle (fig. 5). The lithosphere is broken up into seven large plates and many smaller ones (fig. 6). Some of the plates consist primarily of oceanic crust, others are largely continental, and some are a combination of both. The plates form a giant jigsaw puzzle. They are constantly in motion, riding on the asthenosphere, the hot, plastic layer of the upper mantle. As the

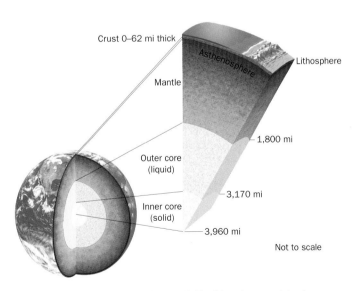

Figure 5. Internal structure of the earth. The lithosphere consists of the crust and solid uppermost part of the mantle overlying the asthenosphere.

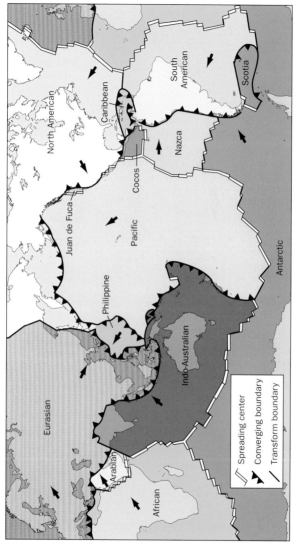

Figure 6. Tectonic plates. Arrows indicate direction of plate movement.

Spreading center
Converging boundary
Transform boundary

North American

Caribbean

South American

Scotia

Nazca

Cocos

Juan de Fuca

Pacific

Philippine

Antarctic

Indo-Australian

Eurasian

Arabian

African

plates interact, they produce earthquakes and volcanoes, push up mountains, and move pieces of the crust hundreds to thousands of miles.

Present-day movements along the margin between the Pacific and North American Plates play a major role in forming today's landscape. Tectonic activity bends and breaks the crust into hills and valleys along the many faults that divide the Bay Area, re-arranging the rocks in the process. The East Bay Hills, Mount Diablo, and the Santa Cruz Mountains, each the product of tectonic movement, are not more than a few million years old, and these geologic infants are continuing to grow. It is not easy to think of our landscape as young, and even harder to realize that it is changing under our feet. Though we get only infrequent reminders from earthquakes, we should not forget that we live in an area that is highly active geologically.

The Dance of the Plates

The plates are restless and in constant movement: they separate and spread apart, they converge (collide), or they slide past each other (fig. 7). All three dance steps are usually so slow, in human terms, that we are unaware of change. It is only in an earthquake that we are forcefully reminded that the rocks beneath us are not as unmoving as they seem.

Plates move apart at divergent plate boundaries, or spreading centers, where molten rock, called magma, rises from the mantle to create new oceanic crust (fig. 8). Spreading centers form a worldwide chain of underwater volcanoes and volcanic mountains, like the one in the middle of the south Atlantic Ocean, separating the African and South American Plates.

Plates collide at convergent plate boundaries, which are the most dynamic geologic settings on earth. Large earthquakes and high volcanic mountain ranges, such as the Andes, the Aleutian Range, and the Cascade Range, occur at these boundaries. At convergent margins, also called subduction zones, commonly the denser plate sinks beneath the other in a process called subduction. Where the plates are an oceanic and a continental plate, the denser oceanic plate subducts beneath the lighter continental plate (fig. 9). Where two continental plates collide, neither sub-

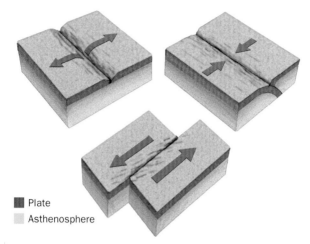

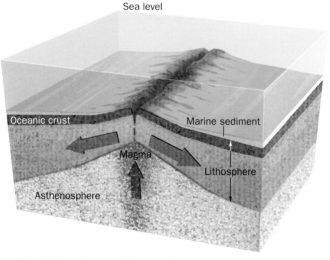

Figure 7. The three types of movement between plates: *upper left*, diverging (spreading); *upper right*, converging (colliding); *lower*, transform (sliding).

Sea level

Oceanic crust

Marine sediment

Magma

Lithosphere

Asthenosphere

Plate

Asthenosphere

Figure 8. Spreading center (divergent plate boundary), where magma rises to form new basaltic oceanic crust, which cools and sinks as it moves away from the spreading center.

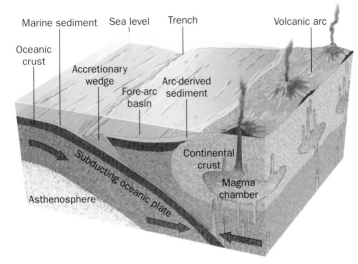

Figure 9. Simplified model of subduction between an oceanic plate and a continental plate.

ducts, and the plate margins are folded and faulted into high regions such as the Himalayas and Tibetan Plateau, where the Eurasian and Indo-Australian Plates are colliding. Technically, geologists use the term "plate collision" only for the latter case, that is, when two continental plates converge, but in this book, it is used in the more general sense of convergence between any two plates.

During subduction, the oceanic crust of the sinking plate, with its thin layer of marine sediment, is carried ever deeper into the earth, gradually heating as it goes. Eventually it begins to melt, forming magma, which rises through the rocks of the upper plate toward the surface, because the melted material is less dense (hence more buoyant) than the colder rock around it. The magma accumulates in magma chambers within the earth's crust and may rise all the way to the surface, where it erupts and forms a chain of volcanoes called a volcanic arc.

When subduction ends, the volcanoes stop erupting, and the magma chamber miles below the surface cools slowly and crystallizes, partly as granitic rock. Over millions of years, the miles of overlying rock may erode away, eventually exposing the crystallized magma chamber at the surface. Around the world granitic rocks are one of the signatures of past subduction. In the Sierra Nevada we can hike over granitic rocks that formed in magma chambers about 100 million years ago and miles below the earth's surface.

At a sliding plate boundary, plates slide past each other along major faults called strike-slip faults. In the Bay Area we live along this type of plate margin. The San Andreas Fault System is a zone of mostly strike-slip faults along which the Pacific Plate is sliding past the North American Plate.

California's Dynamic Past

About 145 million years ago, in the Mesozoic Era, the eastward-moving Farallon Plate (the predecessor of today's Pacific Plate) began to collide with the North American Plate. For almost 100 million years, the two plates collided, and the Farallon Plate was subducted beneath the North American Plate. Large magma chambers formed in the crust many miles below the surface as the subducting plate began to melt. Magma rose to the surface and formed a chain of volcanoes in present-day eastern California.

During this long episode of subduction, many of the Bay Area's most interesting rocks were formed, some locally, others far away (see chapter 3). Today we have rocks in the Bay Area that have traveled thousands of miles on moving tectonic plates—rocks formed halfway around the world. The Farallon Plate brought to California both oceanic crust from a spreading center far to the west and marine sediments deposited on the crust as it traveled. When the Farallon Plate met the North American Plate and began to be subducted, some of the crust and sediments were scraped off onto the North American Plate in a process called accretion. The scraped-off material is called an accretionary wedge or prism (fig. 9). That is how ocean crust and marine sediments from thousands of miles away became part of California. During more than 100 million years of subduction, accretion brought together many of the different types of Mesozoic rocks we see today in the Bay Area.

About 90 million years ago something about the subduction process changed, and no younger granitic rocks are found at the surface in California. Perhaps the angle changed at which the subducted plate was going down, and it did not heat up enough to create large amounts of magma. About 25 or 30 million years ago, the Farallon Plate was almost entirely consumed beneath the North American Plate, and the Pacific Plate first met the North American Plate. Subduction stopped where the latter two plates came into contact, and the Pacific Plate began to slide northward past the North American Plate. Strike-slip movement along the San Andreas Fault System replaced subduction over millions of years. The transition from subduction to sliding along the San Andreas Fault gradually progressed northward and reached the Bay Area about 15 million years ago.

Today subduction is still taking place off the coast of northern California, Oregon, and Washington. This subduction produces occasional large earthquakes and forms the active Cascade Range volcanic chain, which includes Mount Rainier and Mount St. Helens in Washington, as well as Mount Shasta and Lassen Peak in northern California.

The Bay Area's Network of Faults

In central California today the dominant motion between the Pacific and North American Plates is strike-slip (sliding). The Pacific Plate is sliding past the North American Plate at an average rate of about 1.6 inches per year. The boundary between the Pacific and North American Plates is not a sharp one. Although we commonly say that the San Andreas Fault is "the plate boundary," in fact, movement between the plates affects a broad zone of faulting and folding that extends across the entire Bay Area. In the opinion of many geologists, it actually continues a hundred or more miles to the east. For simplicity we refer to the San Andreas Fault as the plate boundary, but remember that the geologic picture is actually much more complicated.

The San Andreas Fault System consists of two major branches in the Bay Area, a western branch consisting of the San Andreas and San Gregorio Faults, and the East Bay Fault System (map 4). The branches join south of San Jose. The San Andreas and San Gregorio Faults run along the coast. In the North Bay, the San Andreas lies partly offshore. The Hayward, Calaveras, and several

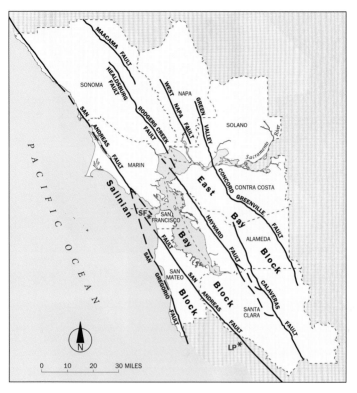

Map 4. Known presently active strands of the San Andreas Fault System in the Bay Area. Note that fault offset is transferred from the Calaveras to the Hayward Fault and from the Hayward to the Rodgers Creek Fault through a region of oblique faults. Also shown are the Bay Area blocks and epicenters of the 1906 San Francisco (SF) and 1989 Loma Prieta (LP) earthquakes (red stars).

other faults make up the East Bay Fault System. Faults to the east include the Rodgers Creek–Healdsburg–Maacama Fault and the Green Valley Fault, which is a continuation of the East Bay Fault System.

The strike-slip movement along the San Andreas Fault System is distributed across all the active faults in the Bay Area. The

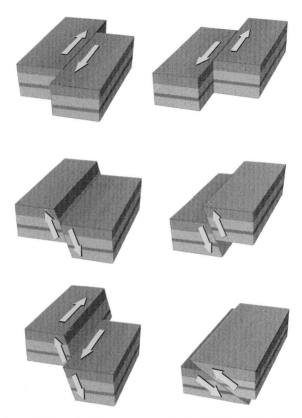

Figure 10. Types of relative movement on faults: right-lateral strike-slip *(upper left);* left-lateral strike-slip *(upper right);* normal *(center left);* reverse *(center right);* oblique-slip *(lower left);* low-angle reverse (thrust) *(lower right).*

movement is right-lateral; that is, when you stand on one side of the fault (either side), the other side appears to move to your right (fig. 10, *upper left*). For example, if you are standing on Mount Tamalpais looking northwest across the San Andreas Fault, Point Reyes on the Pacific Plate is sliding slowly to your right. In the other direction, standing on Point Reyes, the relative motion of Mount Tamalpais is also to your right, although, in

Figure 11. Bay Area seismicity, 1972 to 1989. Yellow dots indicate earthquakes of magnitude 2 or greater.

fact, the Pacific Plate is moving northwestward. Movement along faults can be vertical, horizontal, oblique, or a combination. The faults of the San Andreas Fault System are mainly strike-slip, but many have a component of vertical motion.

Earthquakes are triggered by the release of energy when a fault ruptures. They occur on the Bay Area's many active faults every day (fig. 11), but only a few are large enough to be felt. Strong earthquakes take place where sections of a fault have been stuck, or "locked," and stress energy builds up until it overcomes the strength of the rocks. Sections of faults such as the San Andreas may be locked for hundreds of years, resulting in very large earthquakes like that in 1906, when rocks west of the fault slid past the eastern rocks as much as 21 feet. In the 1989 Loma Prieta earthquake, the horizontal movement was about 6 feet, although

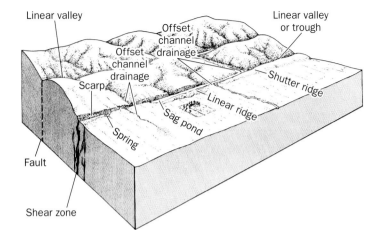

Linear valley

Offset channel drainage

Offset channel drainage

Linear valley or trough

Shutter ridge

Scarp

Linear ridge

Sag pond

Spring

Linear ridge

Fault

Shear zone

Figure 12. Geomorphic expression of fault movement.

the fault didn't break to the surface. The epicenters (sites of first rupture) of the 1906 and Loma Prieta earthquakes are shown on map 4. The 1906 epicenter was originally placed in Olema, in Marin County, where the maximum displacement took place. Recently, the epicenter was relocated to west of San Francisco after a new study of earthquake records.

The amount of energy released by an earthquake is measured by magnitude and usually has to be magnitude 2 to 3 before we feel it. The 1906 San Francisco earthquake, which occurred before instruments were available to measure an earthquake's energy, is estimated to have been magnitude 7.8. In older literature, it was listed as magnitude 8.3, but estimated magnitudes of early earthquakes were recently recalculated based on a better understanding of how earthquake waves travel and what effect they have. There have been nine other earthquakes of magnitude 6.5 or greater in the Bay Area since 1800, including the 1989 Loma Prieta earthquake of magnitude 6.9.

Another type of fault movement is creep, a slow horizontal movement of one side of a fault past the other. Creep does not result in earthquakes. Some faults move by both earthquake and creep, some just by one or the other. The Hayward Fault is creeping at a rate of almost one-quarter inch per year along much of its

Plate 11. The San Andreas Fault valley on the Peninsula. View to west across Upper Crystal Springs Reservoir.

length. The results can be seen especially clearly in the city of Hayward, where sidewalks and curbs have been offset (pl. 102). The Calaveras Fault is creeping at a rate of about one-half inch per year south of the Bay Area, and the San Andreas Fault up to 1.25 inch per year south of San Juan Bautista in San Benito County. Why creep occurs along some parts of the San Andreas Fault System but not others is not clear. Some studies suggest that a fault creeps where it cuts through serpentinite, a rather slippery rock. But the Hayward Fault creeps where no serpentinite has been detected; therefore, that cannot be the only reason, and investigations are continuing.

Movement along faults over a long period of time produces recognizable features in the landscape (fig. 12). Rock along a fault may be ground up, making it more easily eroded to form valleys. The long linear valley of the San Andreas Fault down the Peninsula is a fine example (pl. 11). Streams offset by the Hayward Fault can be seen in relatively undeveloped areas between Hayward and Fremont. They are also evident in the offset of Strawberry Creek and neighboring streams in Berkeley (pl. 12). San Andreas Lake and Crystal Springs Reservoir on the Peninsula, Lake Temescal in Oakland, and Lake Elizabeth in Fremont are all sag ponds, water-filled depressions along the fault, that have been dammed for reservoirs or recreation.

Earthquakes that occurred before records were kept can be

Plate 12. Offset of Strawberry Creek, University of California at Berkeley campus, looking east from Campanile. Strawberry Creek (at arrows) flows through Strawberry Canyon to Memorial Stadium (right center), then north (to left) for about 1,000 feet along the Hayward Fault, then west again through the campus to left of buildings at lower right.

Plate 13. Trench across Concord Fault in Concord, Contra Costa County. Note BART tracks in background.

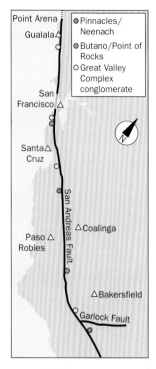

Figure 13. Offsets across the San Andreas Fault Zone. Note that Great Valley Complex conglomerates occur at three locations west of the fault.

studied by digging a trench across a fault to analyze the displacement of rock or soil (pl. 13). By dating charcoal or other materials in the soil, geologists can determine the age of the displacement and can reconstruct the dates of past earthquakes on that fault. From these dates we can determine the recurrence interval, the time between earthquakes on a fault, and estimate how long it might be before another large earthquake occurs. Periodically, seismologists assess the likelihood of a significant earthquake along the various components of the San Andreas Fault System in the Bay Area. This information is available on the U.S. Geological Survey Web site (see Further Reading). Most Bay Area communities have staff to assist residents with personal and neighborhood earthquake preparedness. An earthquake could take place at any time. Take advantage of these resources and be prepared!

In order to tell how much movement has occurred along faults in the past, geologists match distinctive rock formations on either side of the fault, rocks that must originally have formed at the same time and location. The age of the rocks and the distance that today separates the two parts tell us how far they have been offset by fault movement. For example, the Pinnacles (pl. 93), a distinctive volcanic formation west of the San Andreas Fault and just south of the Bay Area in San Benito County, were formed in southern California about 23 million years ago. The distinctive Pinnacles rocks match the

Neenach Volcanics, which lie east of the San Andreas Fault in the Mojave Desert (fig. 13). The San Andreas Fault has carried the Pinnacles 192 miles to their present location in the past 23 million years. Another important match is that of the Butano Sandstone on the Peninsula west of the San Andreas with the Point of Rocks Sandstone in southern California east of the fault, an offset of 225 miles. The oldest rocks so far matched across the plate boundary are sedimentary rocks of Cretaceous age (75 million years old) now present on the Mendocino coast south of Gualala. These rocks, which include a conglomerate with distinctive molluskan fossils, are composed of material eroded from rocks at Eagle Rest Peak 350 miles southeast on the other side of the San Andreas Fault. As fault movement has moved pieces of this rock northward, bits have been left behind at several locations, including a sliver on the Peninsula (fig. 13).

Squeezing the Bay Area

Although the dominant movement between the Pacific and North American Plates now is strike-slip, about 3.5 million years ago or perhaps several million years earlier, a small change in plate motion took place. The Pacific and North American Plates began to collide at a slight angle instead of just sliding past each other. This combination of two types of fault movement is called transpression. Much of the hilly Bay Area landscape has been formed in the past several million years by this squeeze play between the plates. Today 90 percent or more of the movement between the plates is strike-slip and 10 percent or less is compressional. Mount Diablo and the East Bay Hills began to rise just 1 or 2 million years ago and continue to rise about one-sixteenth inch (1 to 2 millimeters) per year. In the South Bay the Santa Cruz Mountains are going up on the west and the Diablo Range on the east. Mount Tamalpais and the mountains in the North Bay are also young and rising.

Transpression compresses and thickens the crust in two ways: by folding and by faulting (fig. 14). Some of the Bay Area hills and valleys are large folds or wrinkles in the crust, like the folds in a tablecloth if you push the edges together. A convex upward fold is called an anticline; a concave fold is a syncline. Thrust faults, a

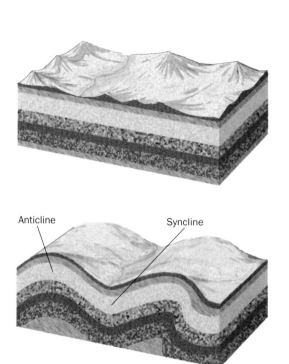

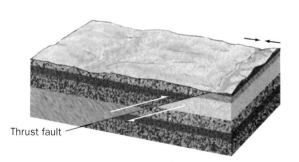

Figure 14. Styles of deformation in the Bay Area: *top,* undeformed rock layers; *center,* folded rock layers, showing anticline and syncline; *bottom,* thrust fault. White arrows show direction of movement; black arrows show direction of compressional stress.

kind of fault in which one block slides up and over another at a low angle, are common around the Bay Area. Parts of the Santa Cruz Mountains are moving up along thrust faults, as are Mount Diablo and Mount Tamalpais. Transpression has shortened the crust in a northeast-southwest direction so that today Mount Diablo, for example, is closer to San Francisco than it was 1 million years ago.

Crustal folding is intensified where a fault bends or is offset, like the slight bend made by the San Andreas Fault southwest of San Jose or the offset between the Hayward and Mission Faults near Fremont. Local uplift or down-dropping of the crust commonly results, and depending on the fault geometry, hills or val-

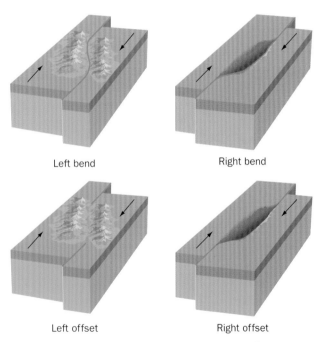

Left bend   Right bend

Left offset   Right offset

Figure 15. Uplift and down-dropping caused by bends and offsets along the right-lateral strike-slip faults of the San Andreas Fault System: a left bend *(upper left)* or left offset *(lower left)* forms uplifted hills; a right bend *(upper right)* or a right offset *(lower right)* forms down-dropped basins or valleys. Fault shown in red.

leys are formed. Where a right-lateral strike-slip fault, like the San Andreas, is offset or bends to the left, the result is compression and uplift, often involving thrust faults (fig. 15). It is not accidental that Loma Prieta and Mount Umunhum near the bend in the San Andreas Fault in the Santa Cruz Mountains and Mission Peak at the Hayward–Mission Fault offset near Fremont are the highest parts of their ranges. Where a right-lateral strike-slip fault is offset or bends right, the result is extension and down-dropped areas forming valleys. Much of the present hilly topography of the Bay Area is related to thrust faulting along offsets or bends in faults.

The Bay Area's Complex Structure

Past plate movements and more recent movement along the San Andreas Fault System have sliced the Bay Area into a great many packets of rock that range in size from large units of many square miles to small local units that are not recognizable without a geologic map. Geologists use a number of terms in referring to the packets of rock, based on their size, age, rock components, geologic history, or a combination of these factors. The most common terms are "block" and "terrane." In general, a block is a large unit, like the San Francisco Bay Block (fig. 16) between the San Andreas and Hayward Faults (map 4). The term "terrane" usually refers to packets of rocks involved in the dynamic plate movements of the Mesozoic and early Cenozoic; they are described in more detail in chapter 3. Unfortunately, usage is not precise, and the same term may be used in several different senses by different geologists.

Each packet, whether large block or small sliver, is bounded by faults. It may consist of rocks totally different from those in neighboring packets, or it may differ from its neighbors in only minor constituents. The individual packets have moved as a unit along the bounding faults. Thus, throughout the Bay Area very different rocks have been rearranged and juxtaposed over time. The Bay Area geologic mosaic is composed of packets of rocks that have traveled from somewhere else to their present position. Some have come great distances, others just a short way. Only the youngest rocks and sediments are truly local, that is, formed near their present location.

Today's major active faults divide the Bay Area into three large blocks: the Salinian Block, west of the San Andreas Fault; the San Francisco Bay Block (fig. 16), between the San Andreas and Hayward Faults; and the East Bay Block, east of the Hayward Fault (map 4). The Salinian Block is moving slowly northwestward toward Alaska with the Pacific Plate. Coastal San Mateo County, most of the Santa Cruz Mountains, Point Reyes, and Bodega Head are going along for the ride. The Salinian Block is a temporary visitor to the Bay Area. Return to the Bay Area in your time machine 2 million years from now, and at the present rate of motion Point Reyes will be 50 miles or so farther north, near the Sonoma-Mendocino county line.

The Bay Block and the East Bay Block are on the North American Plate; however, they, too, are slowly moving along strike-slip faults. If you took a time-lapse photograph of the Bay Area from space over a million years, you would see each block shifting northwestward relative to the block to its east, like dominoes sliding past each other (fig. 16). Remember that the actual boundary between the North American and Pacific Plates is a very broad zone of faulting and movement, over a hundred miles wide. The western edge of the North American Plate is being dragged along as the Pacific Plate slides by.

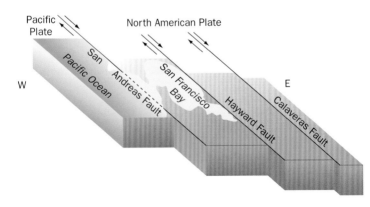

Figure 16. Relative movement of fault slivers. Each block moves northwestward relative to the block to its east. Arrows indicate relative motion.

More than 140 million years of plate tectonic movements have created the Bay Area's geologic mosaic. If you become familiar with the broad pattern of active faults in the Bay Area and how they shape the hills and valleys and rearrange the rocks, you will understand a great deal about the landscape of the Bay Area—and be better prepared for the inevitable earthquakes we experience.

THE WORLD-FAMOUS BAY AREA rocks tell a geologic story that reads like a Russian novel with a very large cast of characters. Because of our plate tectonic history, we have a crazy-quilt pattern of rocks almost defying description and order. Bay Area rocks are not piled up in a tidy sequence of units, like the Grand Canyon's "layer-cake" geology, with older rocks below and younger above. The Bay Area's rock complexity is the result of movement along ancient and young faults over millions of years. A glance at the geologic map (map 2) shows you what an extraordinary mosaic of different rock types occurs here. They are famous for the story they tell and the dramatic Bay Area settings in which you can see them. Even more important, study of these rocks made critical contributions to the development of plate tectonic concepts.

These rocks are the clues, the evidence, by which the story of the Bay Area's landscape and its geologic history can be read. They are windows into the past and give us important information about the geologic environment in which they were formed. Those that can be dated by radiometric techniques or fossils provide a framework for geologic events over time. For over 100 years, Bay Area geologic detectives have been analyzing these rocks. We now have a good understanding of the framework, but many questions remain unanswered.

All three basic types of rocks—igneous, sedimentary, and metamorphic (table 1)—are present in the Bay Area in great variety. Of the igneous rocks, volcanics are found throughout the region, but plutonics are not common. They occur largely at scattered locations west of the San Andreas Fault. Sedimentary and metamorphic rocks are found on both sides of the fault. Some of the more common Bay Area rock types are illustrated in pls. 14a–h.

Rocks formed at about the same time and place and in the same geologic environment are called a formation. A number of different types of rocks related by tectonic events may be referred to as a complex, for example, the Franciscan Complex described below. Although you do not have to recognize all the different types of Bay Area rocks to understand the landscape, getting to know the more significant and interesting rocks will greatly increase your enjoyment and understanding of the regional geology. A description of them all would be a book in itself, but the following summary will introduce you to the major types, and each regional chapter will say more about the most significant rocks of that area.

TABLE 1. Types of Rocks

Type	Formation Process	Examples
Igneous	Formed as molten rock (magma) cools	
Plutonic	From magma that cools underground	Granite, granodiorite, diorite
Volcanic	From magma that erupts and cools at the surface	Obsidian (volcanic glass); lavas such as rhyolite, andesite, basalt; tuff
Sedimentary	Formed at the earth's surface	
	From fragments of other rocks	Sandstone, graywacke, shale, conglomerate
	From organic material, such as skeletons of plants and animals	Limestone, radiolarian chert
	From chemical precipitation	Salt
Metamorphic	Formed by the alteration of rocks of any kind by heat, pressure, or both	Schist, gneiss, marble, slate, serpentinite

The Foundation under Our Feet: The "Basement" Rocks

Let us begin with the older rocks, the basement rocks that form the underpinnings of the Bay Area. They date from the time of the dynamic Mesozoic plate collisions described in chapter 2. They are called the Franciscan, Great Valley, and Salinian Complexes, and they tell the story of subduction and fault movement along the West Coast for over 100 million years. The three complexes overlap in age, but they are different types of rock, and each was formed in a different geographic area and in a different geologic setting. They have been brought together in the Bay Area, some from thousands of miles away or from deep in the earth, by plate motion and faulting. One of the major detective stories of Bay Area geology has been to decipher how these rocks were formed, where they came from, and how they got here.

text continues on page 55

Plate 14a. Pillow basalt (igneous), Franciscan Complex, Cronkhite Beach, Marin County.

Plate 14b. Basalt (igneous), Moraga Formation, East Bay Hills.

Plate 14c. Graywacke (sedimentary), Franciscan Complex, Telegraph Hill, San Francisco.

Plate 14d. Sandstone (sedimentary), Orinda Formation, East Bay Hills.

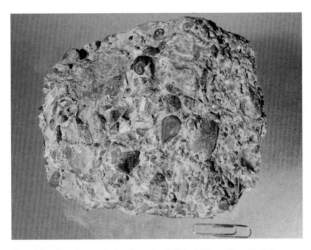

Plate 14e. Conglomerate (sedimentary), Orinda Formation, East Bay Hills.

Plate 14f. Chert (sedimentary), Claremont Formation, East Bay Hills.

Plate 14g. Radiolarian chert (sedimentary), Franciscan Complex, Marin Headlands.

Plate 14h. Serpentinite (metamorphic), Fort Point, San Francisco.

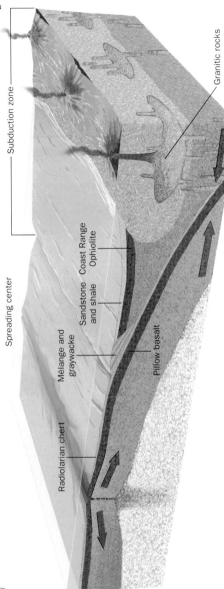

Figure 17. Simplified model of plate tectonic setting in which the types of rocks that make up the Franciscan, Great Valley, and Salinian Complexes formed. Franciscan Complex rocks include pillow basalt (ocean crust), radiolarian chert (marine sediment), graywacke and shale (sediments eroded from land), and mélange. Great Valley Complex rocks include sandstone, shale, and conglomerate eroded from the volcanic arc and continent, and Coast Range Ophiolite (mantle and ocean crust). Salinian Complex granitic rocks crystallized in magma chambers. *Note:* This is a model of how the rocks formed, not a geographic representation of their relationships. Remember that they originated in different geographic areas and were brought together in the Bay Area by fault movements.

The World-Famous Franciscan Complex

Many geologists would say that the Franciscan Complex rocks, named for San Francisco and known as "the Franciscan" for short, are the most interesting of all Bay Area rocks. Geologists come from all over the world to study them. They were formed in several plate tectonic settings (fig. 17), many far from their present location and rafted to California on moving plates. The most common types are described below, others in the regional chapters.

Pillow Basalt

The base of the Franciscan Complex in many areas consists of the upper part of ancient ocean crust composed of basaltic lava flows and pillow lava, also called pillow basalt (pls. 14a, 15). These flows occurred about 100 to 200 million years ago in two geologic settings: at a spreading center thousands of miles to the west or at an oceanic plateau or seamount formed at or near a hot spot. When lava erupts into the ocean, it chills and the outside hardens. The inside remains soft, and the lava flows in long tubes whose ends look like pillows. As the lava interacts with seawater, it is slightly metamorphosed and develops a greenish color (from chlorite and other iron-bearing minerals). It is then called green-

Plate 15. Franciscan pillow basalt near Nicasio Reservoir, Marin County.

Plate 16. Franciscan greenstone at Cronkhite Beach, Marin County.

stone (pl. 16). About 10 percent of the Franciscan Complex consists of pillow lava and greenstone. Today you can see excellent exposures in the Marin Headlands, near Nicasio Reservoir in Marin County, and at Mount Diablo.

Radiolarian Chert

The marine sediment that was deposited on top of the ocean crust basalt is largely composed of the skeletons of microscopic single-celled animals called radiolaria (fig. 18). These organisms live primarily near the surface of the sea and make their skeletons of the mineral silica. As they die, the skeletons slowly drift down to the ocean floor. Because the skeletons accumulate far from land, little other sediment is added—only a bit of dust and volcanic ash. Over time, the sediment hardens into a reddish brown rock called radiolarian chert (pls. 14g, 17), and the dust and ash form thin layers of shale between the chert layers (pl. 44). The chert layer is often tightly folded, as in excellent exposures in the Marin Headlands, on Twin Peaks in San Francisco, on Mount Diablo, and in Coyote Hills Regional Park near Fremont.

In the Marin Headlands the chert was originally over 250 feet thick (today it is thicker because many small thrust faults have cut the Headlands rocks and piled up several sections of the chert). Imagine the enormous number of individual radiolarian

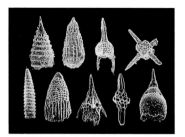

Figure 18. Jurassic radiolarians.

Plate 17. Photographer John Karachewski at folded Franciscan radiolarian chert at Marin Headlands, Marin County. Note tight chevron fold to John's left.

skeletons required to make even 1 inch of this 250-foot sequence and the great span of time it took to accumulate them. The radiolarian skeletons in the Headlands chert began to accumulate in the Pacific Ocean near the equator thousands of miles to the west and about 200 million years ago. But geologic time is long, and the skeletons rained down for 100 million years as the oceanic plate moved slowly eastward to its meeting with North America.

Although chert is typically reddish brown due to its iron content, it may be altered by hot fluids to a multihued rock. Beautiful pebbles of altered chert and other Franciscan rocks make up much of the "sand" on Bay Area beaches north of the Golden Gate (pl. 18).

Plate 18. *Right,* Multicolored Franciscan pebbles from Kirby Cove, Marin County. *Below,* Ridge (foreground) above Kirby Cove. The pebbles eroded from the ridge, were transported to the beach, and were polished by the waves.

Graywacke and Shale

As the Farallon Plate neared the subduction zone and collided with the North American Plate (see chapter 2), great quantities of sediment eroded off the North American Plate and were deposited on top of the chert in a trench that formed at the subduction zone (figs. 9, 17). The sediments hardened into a sandstone called graywacke (pls. 14c, 19), which is gray when fresh but weathers to brown. It is not a "clean" sandstone, like a beach sand that has been sorted by waves to particles of similar size. It is a "dirty" sandstone, which includes a wide range of particle sizes and compositions. The mineral composition of graywacke varies from place to place, reflecting the land from which the sediments were eroded. The Franciscan Complex includes many different kinds of graywacke, formed in widely separated areas and brought together by fault motion.

Plate 19. Franciscan graywacke at Telegraph Hill, San Francisco, in 1976.

The sediment was carried into the trench by turbidity currents (fig. 19), which are massive underwater "landslides" that are a slurry of water and sediment. As a turbidity current flows onto the ocean floor, the coarser sediments settle out first, then finer and finer particles. This process forms graded bedding with the largest particles at the base of each turbidite deposit, grading upward into sandstone (pl. 20), with fine-grained shale at the top. The next turbidity flow deposits coarse material on the shale at the top of the previous flow. The result is a series of alternating layers of sandstone and shale. Turbidite deposits that consist largely of sandstone are formed closer to shore; those composed mostly of shale are formed farther out in the ocean (pl. 105). Good examples of layered Franciscan turbidites can be seen at the east end of the Richmond–San Rafael Bridge.

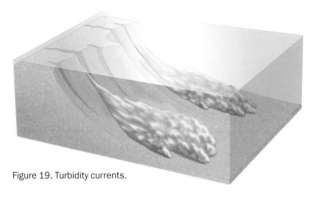

Figure 19. Turbidity currents.

Plate 20. Graded bedding in Pigeon Point Formation at Bean Hollow State Beach, San Mateo County. Franciscan turbidities are similar.

Sometimes such a large amount of sediment flows into the trench at one time that it does not form layers but rather comes to rest as a massive deposit, like the graywacke at Telegraph Hill in San Francisco (pl. 19). The time interval between successive turbidity flows may be tens or hundreds of years, and it is likely that earthquakes are responsible for starting many of them. About 80 percent of the Franciscan consists of graywacke and shale. It has been estimated that at least 350,000 cubic miles of Franciscan graywacke accumulated during subduction—enough to cover the state of California with a layer 10,000 feet thick!

Metamorphics

When Franciscan rocks were subducted during the plate collisions, many were metamorphosed, a process that alters rock through heat or pressure or both. The minerals in the rock are changed from those that were stable under conditions when the rock originally formed to minerals stable under the new conditions. Metamorphism in a subduction zone generally occurs at high pressure but low temperature as rocks are dragged deep into the earth (about 12 to 25 miles) more quickly than they can heat up. In many Franciscan rocks the alteration is not great, and many original minerals remain in the rock. Much of the graywacke so common in the Bay Area has been altered in this way. In other cases, the alteration is more extensive.

The Franciscan Complex includes rare metamorphic rocks formed in the subduction zone under conditions of high pressure and high temperature, called "high-grade" metamorphism. Such rocks are found "floating" in Franciscan mélange in many areas. Under high-grade conditions, which are not common in subduction zones, the original minerals are completely changed to new minerals. In the Franciscan, this alteration results in especially interesting rocks such as eclogite, a greenish rock often

Plate 21. Franciscan eclogite with garnet.

sprinkled with tiny red garnets (pl. 21), and blueschist (see chapter opening photo). These metamorphic rocks are both beautiful and geologically significant. For a long time their occurrence in the Franciscan was an enigma to geologists. Then, as plate tectonic processes became better understood, geologists proposed that these conditions occur during the earliest phases of subduction, when the rocks of the subducting plate first come in contact with hot rocks in the mantle.

These and other Franciscan metamorphics can be seen at Ring Mountain on Tiburon (pl. 47), in Sunol Regional Wilderness south of Pleasanton, in Henry Coe State Park in the South Bay, around Cazadero in the North Bay, and at scattered localities in the Diablo Range.

Our state rock, serpentinite, is a common metamorphic rock throughout the Bay Area. Geologists have long considered it an important component of the Franciscan and Great Valley Complexes. Serpentinite is faulted into the Franciscan Complex at several locations in the Bay Area, but much of it is associated with the Great Valley Complex. It is described further in the section on Great Valley Complex rocks.

Assembling the Franciscan

Although they were formed in different geologic environments, the Franciscan rocks are connected by the plate tectonics that brought them together during the Mesozoic plate collisions. The clue to their association is that at many locales in California we find pillow basalt, chert, and graywacke together, one resting on another without faults between them (fig. 20). Pillow basalts that erupted at a spreading center far to the west under the Pacific Ocean formed the new ocean crust on the Farallon Plate. As the oceanic plate moved slowly eastward, radiolarian skeletons drifted down, grain by grain, onto the ocean crust. The skeletons accumulated and eventually turned to rock, forming chert. The oceanic plate carried the pillow basalt and chert to the subduction zone and into the deep trench, where sediments that eroded off the North American Plate to the east were deposited on the chert to form graywacke. Some of the marine sediments and graywacke were subducted under the North American Plate, metamorphosed, and faulted back to the surface. Some were scraped off as the Farallon Plate descended and were accreted to the North American Plate. Many of the faults so common in Franciscan rocks date from the time of subduction. More recent San Andreas faulting has further disrupted them. Today the rocks are not commonly found in a tidy sequence. At any one place, we may see just one type, or see them all faulted and smeared together. But enough exposures of the original sequence are pre-

Figure 20. Relationship of Franciscan rocks: *bottom,* pillow basalt; *center,* radiolarian chert; *top,* graywacke.

served in California to make clear how the Franciscan rocks were assembled.

During both the Mesozoic plate collisions and later faulting, some of the Franciscan rocks were thoroughly ground up into a mélange, which consists of a soft crushed shale or serpentinite matrix with blocks of other rocks "floating" in the matrix. These blocks may be any of the Franciscan rock types, and they may be small, a few feet in size, or huge, many square miles in area. Mélange is like a geologic chocolate pudding with raisins, nuts, and marshmallows (the resistant blocks) mixed into it. Areas of mélange form a landscape of soft rounded hills with scattered blocks sticking out of the ground (pl. 22). Much of the gently rolling landscape of western Marin and Sonoma Counties is formed on mélange.

The Franciscan rocks (and the other basement rocks described below) have been subdivided into terranes, which are packages of rocks that differ from their neighbors in rock type, sequence, or geologic history. For example, several of the Franciscan terranes are mostly graywacke, the most common Franciscan

Plate 22. Mélange landscape at Nicasio Reservoir, Marin County, with resistant blocks (light colored) embedded in mélange matrix.

rock; others include chert and basalt, but little graywacke. Each terrane is separated from neighboring terranes by faults, and each represents a packet of rocks accreted during the Mesozoic plate collisions. Some of the terranes are very extensive and others are more localized. Commonly, these terranes are surrounded by mélange. In all, 11 Franciscan terranes have been identified in the Bay Area (table 2). As you explore the Franciscan in different parts of the Bay Area, you will become familiar with the many variations in these most interesting rocks.

Great Valley Complex

The second type of basement rock, the Great Valley Complex, also formed in a plate tectonic setting at the time of the Mesozoic plate collisions. These rocks originated in a marine basin, called a "fore-arc basin," between the subduction trench and the volcanic arc (see chapter 2 and fig. 9).

The Great Valley Complex includes two sequences: rocks from the upper mantle and ocean crust, called the Coast Range Ophiolite; and marine sedimentary rocks that eroded from the volcanic arc or continent and were deposited on the ophiolite (fig. 17). These sedimentary rocks are called the Great Valley Sequence (or Great Valley Group). Great Valley Complex rocks form the basement rocks underlying the Sacramento Valley and are exposed primarily in the eastern part of the North Bay (see chapter 10). Slivers of Great Valley Complex rocks have been faulted into other parts of the Bay Area by movement along the San Andreas Fault System. There are major outcrops (exposed rock) in the East Bay Hills from Oakland to Fremont, on the slopes of Mount Diablo, and in the Diablo Range.

Coast Range Ophiolite

The ophiolite, which lies at the base of the Great Valley Complex, includes plutonic rocks of the upper mantle, basaltic volcanic rocks of the ocean crust, rocks transitional between the mantle and crust rocks, and metamorphosed upper mantle rock—California's state rock, the distinctive blue-green serpentinite (pls. 14h, 23). Serpentinite forms through the alteration of rock from the mantle that is faulted up toward the surface during plate collisions. Interaction with ocean waters alters minerals in the mantle rock to a group of serpentine minerals that form the rock serpentinite. In the Bay Area a long belt of serpentinite is found on

TABLE 2. Terranes in the Bay Area

Terrane	Most Common Types of Rocks	Places to See Them
Franciscan Complex		
Alcatraz	Graywacke, shale	Telegraph Hill, Alcatraz
Burnt Hills	Thin-bedded graywacke and shale	Diablo Range
Cazadero	Altered graywacke, blueschist	Cazadero area, west of Healdsburg
Devils Den Canyon	Graywacke	Northeast of Alexander Valley, Sonoma County
Lake Sonoma	Graywacke, minor chert, greenstone	Southwest of Lake Sonoma, Sonoma County
Marin Headlands	Pillow basalt, chert, graywacke	Marin County, Twin Peaks, Mount Diablo, the Peninsula
Nicasio Reservoir	Pillow basalt from oceanic island like Hawaii	Black Mountain, Marin County
Novato Quarry	Thin-bedded graywacke and shale	Point San Pablo, China Camp State Park
Permanente	Pillow basalt, minor chert, limestone, graywacke	Point Bonita, the Peninsula
San Bruno Mountain	Graywacke, shale	San Bruno Mountain, Bolinas Ridge
Yolla Bolly	Altered graywacke, chert, metamorphics	Angel Island, Tiburon Peninsula, Diablo Range
Great Valley Complex		
Del Puerto	Oceanic crust, silicic sediments, volcanics, marine turbidites	East Bay Hills, Diablo Range
Elder Creek	Oceanic crust, serpentinite, marine turbidites	Vaca Mountains
Healdsburg	Oceanic crust and marine turbidites, including conglomerate	Sonoma County, Black Point in Marin County
Point Arena	Oceanic crust, marine turbidites, tropical mollusk shells	Sonoma coast from Fort Ross to Point Arena in Mendocino County
Salinian Complex		
Salinia	Plutonic rocks, marine sediments, minor metamorphics	Montara Mountain, Point Reyes, Bodega Head

Plate 23 *(right)*. Serpentinite at east end of Perles Beach, Angel Island State Park, showing resistant knobs in crushed matrix.

Plate 24 *(below)*. Serpentinite outcrop showing rubbly soil surface, Lucas Valley Road, Marin County.

the eastern side of the Coast Ranges, and it is present in many fault slivers elsewhere. Serpentinite forms a rubbly landscape because it typically makes only a thin soil cover (pl. 24). Because the soil is deficient in calcium and potassium and contains magnesium, chromium, and other elements toxic to plants, a restricted flora grows on it. Trees and shrubs such as cypress, ceanothus, manzanita and gray pine (also called foothill pine), and a limited number of wildflowers are commonly found on serpentinite soils. Fine examples of serpentinite can be seen on Mount Tamalpais and Mount Diablo, on the Peninsula along Hwy. 280, in the South Bay near Calero Reservoir and in Almaden Quicksilver County Park, and at many localities in the Coast Ranges.

Great Valley Sequence

Deposited on the ophiolite is a very thick sequence of marine sedimentary rocks of Jurassic and Cretaceous age, the Great Valley Sequence (fig. 17). It consists of sediments eroded from the continent and the volcanic arc, which formed as the sinking plate to the west began to melt. The sediments were deposited in the fore-arc basin. Along the eastern side of the Coast Ranges, these rocks are as much as 30,000 feet thick—almost 6 miles of sediment! The Great Valley Sequence consists of dozens of layers of sandstone, shale, and conglomerate (pl. 105), many deposited by turbidity currents. They commonly contain economically important gas deposits such as those in eastern Contra Costa and Solano Counties.

We know a great deal about the Great Valley Sequence because the rock layers were pushed up and exposed at the surface as the Coast Ranges were uplifted (pl. 25). You can see these tilted layers of sandstone and shale along the northeastern margin of the Bay Area in Napa and Solano Counties. The more erosion-resistant sandstone forms ridges; the softer, more easily eroded shale forms valleys. The best places to see these sedimentary layers are east of Mount Diablo and Lake Berryessa (see chapters 9 and 10).

Plate 25. Uplifted and tilted Great Valley Sequence rock units (foreground) at west edge of Sacramento Valley. Snow Mountain and its neighbors on ridge in background consist of Franciscan rocks.

The Great Valley Sequence in the Bay Area has been divided into four terranes (table 2). Three of them, the Del Puerto, Healdsburg, and Elder Creek Terranes, consist of oceanic crust (the ophiolite) and various types of turbidite rocks. The Del Puerto Terrane also includes silica-rich volcanics, and the Healdsburg Terrane includes a conglomerate with distinctive cobbles. The source of the conglomerate is still unknown, but some evidence suggests it was formed in southern California. The fourth, the Point Arena Terrane, which is found only west of the San Andreas Fault, consists of rocks that are quite different from all the other Great Valley rocks and perhaps should be put in a category of their own. They include not only ocean crust and turbidite rocks, but also a conglomerate at Anchor Bay in Mendocino County whose cobbles match rocks found 350 miles to the south on the east side of the San Andreas Fault. As described in chapter 2, this conglomerate has been used to estimate the amount of offset on the fault.

Salinian Complex: Visitor from Down South

The third type of basement rock in the Bay Area comprises plutonic rocks of the Salinian Complex, also called the Salinia Terrane (table 2). They occur only west of the San Andreas Fault in Marin, Sonoma, and San Mateo Counties and at the Farallon Islands. Like the Franciscan rocks, they were formed elsewhere during the Mesozoic plate collisions. They were brought to the Bay Area by movement on the San Andreas Fault System, and the Bay Area is the northernmost place where Salinian rocks occur.

Salinian Complex rocks also formed in a subduction setting. They include plutonic rocks formed in magma chambers far below the surface as the subducting plate partially melted (fig. 17). They have been exposed by erosion of the overlying rocks. In composition, those on Bodega Head, Point Reyes (pl. 26), the Farallon Islands, and Montara Mountain are relatively rich in silica, rocks referred to as "granitic." They are similar in composition to the granitic rocks found in the Sierra Nevada and are of similar age: 80 to 110 million years old.

The Salinian Complex granitics were formed several hundred miles to the south. Geologic evidence suggests that they formed as a southern extension of the Sierra Nevada and later were separated from the Sierra by San Andreas Fault motion and rafted

Plate 26. Granitic rocks at McClures Beach, Point Reyes.

northward with the Pacific Plate. Paleomagnetic studies have suggested an origin even farther to the south, but more recent work does not support this idea. The San Andreas Fault System is responsible for their displacement, and their presence in the Bay Area is just temporary. They are geologic visitors traveling with the Pacific Plate as it moves past us toward Alaska.

Older Metamorphic Rocks

At a few locales in the Bay Area, small patches of older metamorphosed rocks occur with the granitic rocks. They are remnants of continental crust rocks altered by hot magma that formed as the subducting plate melted and the magma rose into the crust. This type of metamorphism alters the rocks through high heat and high pressure. Most of these older rocks were originally sedimentary rocks such as sandstone, shale, and limestone, now metamorphosed to quartzite, schist, and marble, respectively. The age of the metamorphics is not known. They must be older than the granitic rocks that altered them, and they may be the oldest rocks in the Bay Area. Rocks similar to the original sedimentary rocks occur in eastern California, the Mojave Desert, and northern Mexico, where they are Paleozoic in age (probably about 350 to

450 million years old). It is likely that the metamorphic rocks of the Salinian Block are related to those in southern California.

You have to look carefully for these oldest rocks because not many have survived transport and erosion. The best and most accessible exposures are on Point Reyes, where they occur with the granitics that altered them (pls. 34 and 35). There are also small remnants on Bodega Head and Montara Mountain. Visit them and pay your respects to these ancients of the Bay Area's rocky foundation.

The Younger Rocks: Emerging from the Sea

Resting on the rocks of the three basement complexes are younger sedimentary and volcanic rocks that provide clues to geologic events in the Bay Area over the past 65 million years, the Cenozoic Era. These rocks record two profound changes in the geologic setting: first the change from a colliding to a sliding plate boundary, and then the change from a marine to a terrestrial environment.

At the end of the Mesozoic, 65 million years ago, the continental shoreline was to the east along the present Sierra foothills, and the Bay Area was still covered by ocean waters. As subduction continued into the Tertiary Period (to about 25 million years ago), local faulting broke up the extensive trench in which sediments had been accumulating. More restricted regional marine basins formed, each with its own characteristic sedimentary fill. At times a drop in sea level or local tectonic movements formed shallow marine or terrestrial environments in place of deep ocean basins. Many varieties of sediment—future sandstone, shale, and chert—accumulated in these basins.

The change from subduction to a sliding plate boundary that began about 25 million years ago and the gradual change to a terrestrial environment are recorded in the rocks of the late Tertiary Period (1.8 million to 23 million years ago). Marine sedimentary rocks continued to be deposited in local basins as they became gradually more shallow. One of the most widespread of these rocks is the Monterey Formation (pl. 27), named for its prominent outcrops on the Monterey Peninsula. As a major petroleum

Plate 27. Monterey Formation at Kehoe Beach, Point Reyes.

source rock, it is one of the most important rocks in the central California Coast Ranges. Microscopic fossils of single-celled plants (diatoms) and animals (foraminifers) are common in the Monterey Formation. A similar rock is the Claremont Formation found in the East Bay (pls. 14f, 117). Both are easily recognizable by their thin layering and because they weather to a light color. Dozens of other sedimentary rocks in the Bay Area represent deposition in late Tertiary local marine basins, both deep and shallow. The more important or interesting units are discussed in the regional chapters.

The transition from a marine to a terrestrial environment is best preserved in the sedimentary rocks of the East Bay (see chapter 9). Rock sequences at several sites there mark the change from deep ocean to shallower marine basins, then to estuarine, and finally to terrestrial conditions over a period of a few million years.

Tertiary Volcanic Rocks

Although young volcanic rocks are not common in the Bay Area as a whole, in the North Bay they make up a significant part of the landscape (map 2, pl. 28). Smaller outcrops of young volcanic rocks occur in the East Bay Hills and on the Peninsula. Several

Plate 28. Volcanic rocks (Sonoma Volcanics) along ridge east of Napa Valley.

types are present in the Bay Area, including lava flows, volcanic mudflow rocks, and tuff, which consists of solid fragments of volcanic ash, crystals, and bits of cooled lava (pl. 29). The lavas vary in chemical composition according to the proportion of silica in the rock. Basalts (pl. 14b) are low in silica and generally dark; andesites are intermediate in silica and color; rhyolites are silica rich and light in color.

Volcanic rocks play an important role in deciphering the geologic history of an area because they represent an instant of geologic time (the volcanic eruption that created them), and it is usually possible to date them by radiometric methods. Each eruption has a specific chemical fingerprint, which can be analyzed from the minerals in the rock. Thus, ash deposits that have settled in widely scattered locations can be traced to a specific volcanic eruption, for example, one in the Lassen Peak area. Two volcanic eruptions, in particular, have left important deposits in the Bay Area: a huge eruption in eastern California about 738,000 years ago that deposited the Bishop Tuff, and an eruption from a volcano near Lassen Peak about 570,000 years ago that deposited the Rockland Ash. These ashes have been found in sedimentary deposits in the sea cliffs south of San Francisco, in South Bay de-

Plate 29. Close-up of volcanic tuff, Sonoma County.

posits, and in mud beneath San Francisco Bay. They were carried to the Bay Area by the Sacramento–San Joaquin River system that drains much of California.

The Bay Area volcanic rocks were produced mainly by local eruptions from small volcanoes and vents, not by high volcanoes associated with colliding plates, such as those of the Cascade Range. Many, but not all, Bay Area volcanics are related to the change that started about 25 to 30 million years ago from a colliding to a sliding plate boundary and the beginning of the San Andreas Fault System. This change left a trail of volcanic rocks, older to the south and younger to the north. The Quien Sabe Volcanics in the Diablo Range are 11 to 13 million years old, the volcanics of the East Bay Hills 9 to 10.5 million years old, the Sonoma Volcanics 2.6 to 8 million years old, and the Clear Lake Volcanics

2 to 10,000 million years old. These volcanics are often associated with local faults because fractures in the crust permit magma to rise to the surface. Other Bay Area volcanic rocks, such as the young volcanics near Coyote Reservoir in the South Bay, are not part of this sequence, though they also erupted along a fault. Not all of these volcanics are in the place where they formed. Some have been moved by activity on the East Bay Fault System.

At the Top of the Heap

Lying on top of the Tertiary volcanic and sedimentary rocks are younger sediments, deposited in the past 1.8 million years, that is, in the Quaternary Period. The Bay Area's valleys are filled with materials eroded out of the rising hills, and creek deposits form the sloping alluvial plain that surrounds the bay. Coastal marine terraces are covered with beach and dune sands. Young deposits also occur along the major faults, where local uplifts have led to erosion of the rising land by streams and deposition of the debris in fault valleys. These young deposits are unconsolidated or poorly consolidated; that is, not well compacted or cemented into rock.

At the margin of San Francisco Bay and beneath the bay are thick deposits of sediment carried in by the Sacramento–San Joaquin River system and by local rivers and creeks. Sand and mud have accumulated in the valley of San Francisco Bay for the past half a million to a million years (see chapter 6).

All together the Bay Area's fascinating rocks tell the geologic history of this area. Each part of the Bay Area illustrates a different part of the geologic story. Marin County sheds particular light on various aspects of Mesozoic plate collision and subduction. The Peninsula, South Bay, and East Bay tell us much about past and present movements of the San Andreas Fault System. In the North Bay volcanoes are a key focus, and San Francisco Bay itself provides clues to more recent events. Geologic detective work, using the Bay Area's diverse rocks as evidence, gives us a framework for understanding the Bay Area's varied and complex geology.

THE GEOLOGY OF MARIN COUNTY is remarkable for its variety and significance. Here, within a few hundred square miles, we can decipher the dramatic plate tectonic history of the Bay Area and observe the recent activity of one of the world's major faults, the San Andreas Fault. We can learn about past geologic events and the processes that are actively shaping the landscape today. Marin County (map 5) provides us an opportunity to explore a wide range of geologic processes, rock types, and scenery in a magnificent setting.

The story told by the rocks of Marin County (map 6) spans some 200 million years, with hints from even older times. It has interested geologists from around the world, because rocks that are important to an understanding of plate tectonics are especially well exposed in Marin County. Most of the common rock types associated with the Mesozoic collision between the Farallon Plate (the predecessor of the Pacific Plate) and the North American Plate (see chapters 2 and 3) can be found in this small part of the Bay Area.

Marin County is divided into two geologic landscapes by the San Andreas Fault. West of the fault, Point Reyes rides on the Pacific Plate, moving slowly northwestward past the rest of the Bay Area. The basement (older) rocks on Point Reyes are completely different from those of the rest of Marin County. They are mainly Salinian Complex granitic rocks. East of the fault on the North American Plate, the basement rocks include the volcanic, sedimentary, and metamorphic rocks of the Franciscan Complex.

Marin topography is as varied as the geology, from the heights of Mount Tamalpais (2,604 feet) to the marshes along San Pablo Bay. The landscape is shaped by rock type, tectonic forces, weathering and erosion, and the ocean's assault against the land. The crust has been folded and uplifted into ridges composed of rocks resistant to erosion (pl. 30). Both ridges and neighboring valleys commonly parallel the San Andreas Fault. The lower grassy rolling hills of western Marin County are underlain by relatively soft rock that is prone to slumping and landsliding.

One of the most dramatic geologic features of Marin County is the wide valley formed by the San Andreas Fault, separating the Point Reyes Peninsula from the "mainland." The valley is clearly visible from a high altitude (pl. 31). At its northern and southern

text continues on page 84

Plate 30. Northwest-southeast-trending ridges of Marin County, looking northeast from Mount Tamalpais.

Plate 31. Point Reyes Peninsula, looking east. The San Andreas Fault valley is the prominent dark band across the center of the figure. False-color image by NASA.

Map 5. Map of Marin County.

Map 6. General geology of Marin County.

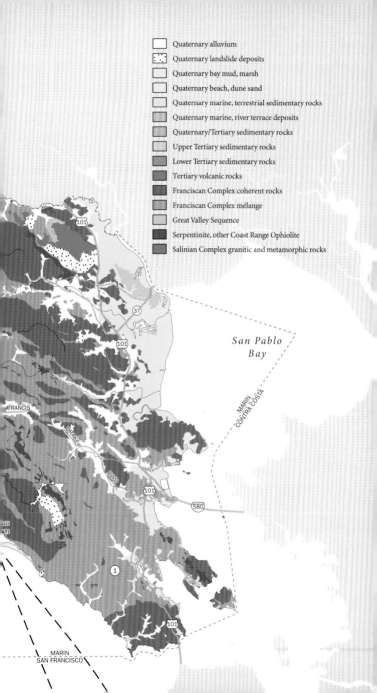

Quaternary alluvium
Quaternary landslide deposits
Quaternary bay mud, marsh
Quaternary beach, dune sand
Quaternary marine, terrestrial sedimentary rocks
Quaternary marine, river terrace deposits
Quaternary/Tertiary sedimentary rocks
Upper Tertiary sedimentary rocks
Lower Tertiary sedimentary rocks
Tertiary volcanic rocks
Franciscan Complex coherent rocks
Franciscan Complex mélange
Great Valley Sequence
Serpentinite, other Coast Range Ophiolite
Salinian Complex granitic and metamorphic rocks

San Pablo Bay

MARIN
CONTRA COSTA

FRANCIS

DRAKE

BLVD

MARIN
SAN FRANCISCO

ends, the valley is flooded by the sea, creating Tomales Bay and Bolinas Lagoon, respectively. Many of the features typical of a fault zone can be seen in this valley (fig. 12). The repeated movement, a few feet or inches in each big earthquake, has rearranged stream drainage, created sag ponds, and crumpled up shutter ridges. Here, we have the strange circumstance of two parallel streams that in places are only one-quarter mile apart but flow in opposite directions: Olema Creek on the east side of the valley drains northward into Tomales Bay; Pine Gulch Creek on the west side drains southward into Bolinas Lagoon. Many small streams draining Bolinas Ridge on the east and some draining Inverness Ridge on the west also have been offset by movement along the fault.

Across from the visitor center at Point Reyes National Seashore headquarters in Olema, a paved earthquake trail introduces you to the effects of the 1906 earthquake. The visitor center is at the site of the former Skinner Ranch, where the rocks to the west moved 14 to 16 feet past the eastern rocks in 1906. Today blue posts mark the ground rupture, and a fence offset has been reconstructed (pl. 32). Rocks along the trail indicate the different rock types on either side of the fault. Nearby, along Bear Valley Road, you can see the sag ponds and shutter ridges typical of a fault zone. Olema was long thought to be the epicenter of the 1906 earthquake, but recent work has relocated the epicenter to an area just offshore of San Francisco.

West of the San Andreas Fault

The Point Reyes Peninsula's landscape is shaped by rocks, sea, wind, and time. Along its seaward edge lies the Point Reyes Beach, 12 miles of crashing waves and shifting sands (pl. 33). Where sea cliffs are low, the sand drifts inland to form dunes, providing shelter from the almost ceaseless wind. The waves move the sand on and off the beach with the seasons and drive it southward by longshore drift (see chapter 1), to the Point Reyes headland, where it is trapped. The sand at Drakes Beach and Limontour Spit along the eastern shore of Drakes Bay (map 7) is derived from the nearby cliffs.

South of Limantour, high rocky cliffs form the Point Reyes

Plate 32. Offset fence along earthquake trail at Bear Valley Visitor Center, Point Reyes National Seashore. View to south. Fence on right has moved toward camera from its original position in left background (by person).

shore. Several remnants of an uplifted marine terrace perch 150 feet or so above the waves. These flat-lying bits of land seem out of place along the rugged shore. Bolinas sits on the largest remaining piece of terrace, "the Mesa," as it is called by locals. This marine terrace was carved at sea level about 80,000 years ago and has been uplifted and tilted in the ongoing dance of the Pacific and North American Plates.

A different process, a drop in sea level, is recorded in the drowned valleys of Drakes Estero, whose fingers extend northward through rolling upland pastures. Oysters are farmed in the inlets, and gulls splash in the warm shallow waters. The estero is a visible reminder of change through time. The inlets are drowned valleys cut into Point Reyes Peninsula by streams flowing off Inverness Ridge during the Ice Age, when sea level was about 400 feet lower than today (see chapter 6). As the glaciers melted and sea level rose, the stream-cut valleys were drowned, creating the estero. Toward Bolinas at the south end of the peninsula, a group of lakes marks yet another geologic process in action, an area of extensive landslides that slid westward off Inverness Ridge and blocked the drainage, ponding the water into lakes.

Plate 33. Point Reyes Beach. View to northeast from road to lighthouse.

Sierra Rocks in the Bay Area?

When you cross the wide, straight valley of the San Andreas Fault and step onto Point Reyes Peninsula from mainland North America, you step into another world, geologically speaking (maps 6, 8). Point Reyes is part of the Salinian Block on the Pacific Plate and is moving northward toward Alaska at an average rate of about 1.6 inches a year (see chapter 2). Granitic rocks 80 to 90 million years old underlie the peninsula and are exposed in the higher hills of Inverness Ridge and at the point itself along the water below its historic lighthouse. At Kehoe Beach several types of granitic rocks are exposed in the cliffs, and they are cut by prominent white quartz veins and dikes (pl. 34). These rocks are highly fractured and faulted, a legacy of their long travels with the San Andreas Fault.

As the granitic rocks were cooling from molten rock deep in the crust (see chapter 3), associated heat and pressure altered the surrounding sedimentary rocks, including limestone, shale, and sandstone, into the equivalent metamorphic rocks, marble, schist, and quartzite. These rocks are probably of mid-Paleozoic age—about 350 to 450 million years old—and may be the oldest rocks in the Bay Area. Most of them have been eroded away, but examples can be found at Teachers Beach on Tomales

Plate 34. Granitic rocks with wide dikes and narrow quartz veins at Kehoe Beach.

Plate 35. Detail of metamorphic boulder of gneiss at south end of McClures Beach. Rock specimen is 3 inches wide.

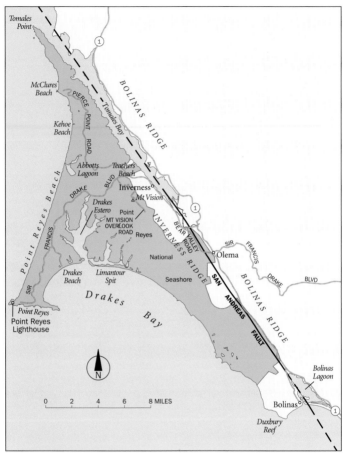

Map 7. Map of Point Reyes.

Bay north of Inverness, along the crest of Inverness Ridge near Mount Vision, and at other scattered localities. An excellent place to see them is at the south end of McClures Beach on Tomales Point, where they are among the large boulders lying at the foot of the cliff (pl. 35). Many of the metamorphic rocks in the cliff are intricately folded and cut by granitic dikes.

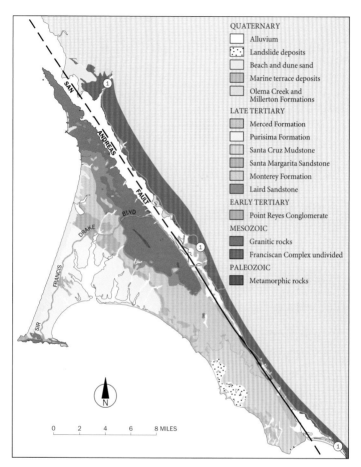

Map 8. Geologic map of Point Reyes.

Within the image legend:

QUATERNARY
- Alluvium
- Landslide deposits
- Beach and dune sand
- Marine terrace deposits
- Olema Creek and Millerton Formations

LATE TERTIARY
- Merced Formation
- Purisima Formation
- Santa Cruz Mudstone
- Santa Margarita Sandstone
- Monterey Formation
- Laird Sandstone

EARLY TERTIARY
- Point Reyes Conglomerate

MESOZOIC
- Granitic rocks
- Franciscan Complex undivided

PALEOZOIC
- Metamorphic rocks

Overlying the granitic rocks is a thick sequence, almost 15,000 feet, of Tertiary sedimentary rocks (maps 6, 8). The oldest are deep marine turbidites, the youngest shallow marine, beach, and dune deposits. Thus, the sequence records the uplift and shallowing of this part of the Salinian Block. Although you cannot see the entire sequence at any one place, each unit is well exposed at

Plate 36. Conglomerate at Point Reyes Lighthouse, showing variety of cobbles, including large, purple volcanic rock (in center).

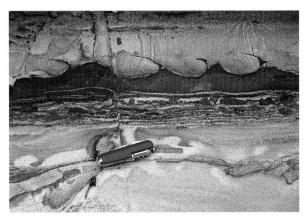

Plate 37. Turbidite structures in the Point Reyes Conglomerate. The wispy fingers at top of dark shale layer are load structures that indicate direction of flow (right to left) of the overlying sandy unit.

some place on Point Reyes. The oldest of these rocks is an Eocene conglomerate that rests on the granitic rocks below the lighthouse. It includes quartzite, large cobbles of the underlying granite, an unusual purple volcanic rock, and others (pl. 36). The pur-

Plate 38. Fault contact between sedimentary Laird Sandstone (right) and granitic rocks (left) at Kehoe Beach.

ple volcanic rocks, which contain small white crystals of the mineral feldspar, are evidence for the long distance these rocks have traveled with the Pacific Plate. The nearest known source of these distinctive rocks is southern California or Mexico.

Several features in the conglomerate indicate that it was deposited by turbidity currents (see chapter 3). The conglomerate is interbedded with sandstones that exhibit features typical of turbidites (pl. 37), and microfossils in the rocks indicate that they were deposited in a deep ocean environment. The conglomerate between the lighthouse visitor center and the stairs to the lighthouse is particularly interesting. Add puffins and cormorants on the rocks far below, sea lions playing in the water, migrating whales in spring and fall, and a beautiful lighthouse with its Fresnel lens intact, and you have one of the finest geologic sites in the Bay Area.

Overlying the conglomerate is a sequence of younger (upper Tertiary) marine sedimentary rocks. The oldest of these is the Miocene Laird Sandstone, which forms interesting cliffs from Kehoe Beach north to its faulted contact with the granitic rocks of Tomales Point (pl. 38). The Monterey Formation, also Miocene

Plate 39. Cliffs of Purisima Formation (formerly called Drakes Bay Formation) at Drakes Beach. Note prominent faults (at arrows) cutting and offsetting sedimentary layers.

in age, lies on top of the Laird Sandstone and can be seen at Kehoe Beach, just north of the sand dune as the trail reaches the beach (pl. 27). Here it is composed of thin, folded layers of white shale. The white color is due to weathering; the fresh rock is black and smells of oil when broken.

On top of the Monterey Formation rest three other marine sedimentary units, the Santa Margarita Sandstone, exposed at Abbotts Lagoon; the Santa Cruz Mudstone with its tilted layers and tidepools at Duxbury Reef near Bolinas; and the Purisima Formation, which makes up the magnificent white cliffs at Drakes Beach (pl. 39). These younger sedimentary rocks were originally given local names, such as Drakes Bay Formation for the rocks at Drakes Beach. A comparison of the Point Reyes rocks with similar rocks in the Santa Cruz Mountains and on the Monterey Peninsula indicates that they are the same rock and were once contiguous. The Point Reyes rocks have been offset about 90 miles by movement along the San Gregorio and San Andreas Faults.

East of the San Andreas Fault

The Franciscan Complex

Turn your back on Point Reyes, go eastward across the San Andreas Fault valley, and you leave granitic rocks behind and enter the far less orderly Franciscan landscape of "mainland" Marin (map 6). The Franciscan Complex is a mosaic of rock types assembled during the long period of Mesozoic plate collisions (see chapters 2 and 3). Most of the typical Franciscan rock types described in chapter 3 can be seen in Marin County, though not all in one place. The most common Franciscan rock, graywacke, is well exposed in numerous road cuts throughout Marin County (pl. 40), along the trails of Mount Tamalpais (see below) and China Camp State Parks, and in McNears Quarry at Point San Pedro, where graywacke has been quarried for construction material since the late 1800s. Along the San Andreas Fault Zone, Bolinas Ridge is graywacke. It makes up the eastern side of the San Andreas Fault valley from Bolinas Lagoon northward almost to Tomales.

Excellent outcrops of Franciscan pillow basalt and radiolarian chert are found in the Marin Headlands (see below). Black

Plate 40. Graywacke (light) and shale (dark) on west side of Hwy. 101, north of Golden Gate Bridge.

Plate 41. Large resistant blocks in mélange in road cut on Hwy. 101 at the top of hill just south of Marin Civic Center (North San Pedro Road exit).

Mountain and its neighbors, west of Nicasio Reservoir, are three huge blocks of volcanic rock. Their pillowed lava flows are beautifully exposed along the road by the reservoir (pl. 15). An especially interesting and important constituent of the Franciscan Complex, the highly unusual blueschist metamorphic rocks, is exposed on Ring Mountain (see below).

The landscape of West Marin consists of soft rolling dairy lands separated by higher northwest-southeast-trending ridges. East of the San Andreas Fault, the long rocky ridges are underlain by several kinds of erosion-resistant Franciscan rocks. The rolling hills owe their sensuous form to Franciscan mélange, which is susceptible to landslides and erosion when it gets water saturated, creating the "melted ice cream topography" for which a mélange landscape is noted. Drive through West Marin when the sun is low in the sky and the shadows show you how wrinkled most of the landscape is. The mélange matrix is sheared graywacke and shale with erosion-resistant blocks of other Franciscan rocks sticking up here and there (pl. 22). The blocks are commonly crowned by a large shrub whose roots seek the moisture retained under the rock. Beware: the shrub is often poison oak *(Toxicodendron diversilobum)*. The resistant rocks can be small blocks a few feet in size or huge blocks many square miles in area. They usually consist of pillow basalt, coherent graywacke, serpentinite, or chert, which is commonly bleached white. Serpentinite also forms blue green outcrops scattered around the mélange landscape. These rolling West Marin hills are the quintessential Franciscan mélange, owing their beauty to the tremendous forces of plate collision in the Mesozoic and the erosional processes of today. Drive through West Marin on any of the pastoral roads that meander across the hills, such as Chileno Valley Road or the Pt. Reyes–Petaluma Road, and you can savor the beauty of a mélange landscape.

Several road cuts along Hwy. 101 in southern Marin County also expose mélange (pl. 41), and in central Marin resistant blocks include some prominent serpentinite and chert outcrops, which can be seen along Hwy. 1 between Mill Valley and Stinson Beach. Evidence of landslides in the easily eroded mélange is particularly clear along this section of road.

Franciscan Terranes

Of the 11 Franciscan terranes found in the Bay Area (table 2), six terranes and mélange are present in Marin County—the Marin Headlands, Yolla Bolly, Novato Quarry, San Bruno Mountain, Nicasio Reservoir, and Permanente Terranes (map 9). Marin is an excellent place to become acquainted with these distinctive packets of rock that were formed at different times and places and then were brought together during the Mesozoic plate collision. The Marin Headlands Terrane is named for the outstanding exposures in the headlands (see below). The Nicasio Reservoir Terrane, which underlies the east slopes of Mount Tamalpais and continues northward to Nicasio Reservoir and Black Mountain, consists mostly of pillow basalt with some chert and graywacke. The Permanente Terrane is found at the tip of Point Bonita and at scattered small outcrops along the San Andreas Fault north of Bolinas.

Three terranes consist mostly of different types of graywacke: the Yolla Bolly Terrane on the Tiburon Peninsula, on Angel Island, and in the northern part of the county; the Novato Quarry Terrane at China Camp State Park and to the northwest along Big Rock Ridge; and the San Bruno Mountain Terrane on the southern part of Bolinas Ridge. All of these terranes are enclosed in mélange, which is an important component of the Franciscan in Marin County.

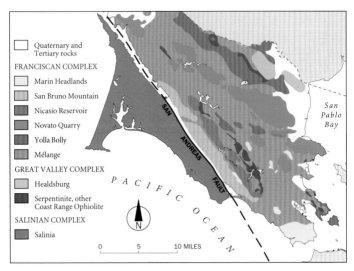

Map 9. Terranes of Marin County. Northernmost part of county not shown.

The Great Valley Complex

Although most of the rocks in "mainland" Marin County are Franciscan, some outcrops of Great Valley Complex rocks are present throughout the county (map 6). Serpentinite, the California state rock and part of the Coast Range Ophiolite, occurs in a discontinuous belt from Mount Tamalpais to Nicasio Reservoir. You can see good exposures around Alpine Lake and at Rock Springs on Mount Tamalpais, along Lucas Valley Road, and on the road between Bolinas and Fairfax (pl. 42). You can recognize the serpentinite by its blue green color and rubbly surface expression (pl. 24). It also occurs as scattered blocks in mélange throughout the county.

Small outcrops of Great Valley Sequence sandstone and shale are present in the Novato area near Burdell Mountain. An interesting conglomerate called the Novato Conglomerate (pl. 43), part of the Healdsburg Terrane, forms the ridge west of the Petaluma River near Black Point. Cobbles in the conglomerate include volcanic, granitic, and chert clasts. Their source may be

Plate 42. Serpentinite along the Bolinas-Fairfax Road.

as far south as the Peninsular Range in southern California, where there are rocks of the type that make up the conglomerate. Similar conglomerates are present in Niles Canyon in Alameda County and at Alum Rock Park east of San Jose (see chapter 8). If all these pieces are the same rock unit, then it has been broken up and spread across 200 miles by fault movement during its trip from southern California to the Bay Area.

The Younger Rocks

In northern Marin County there are sedimentary rocks of Tertiary age that were deposited on top of the subduction-related rocks of the Franciscan and Great Valley Complexes, as well as a variety of volcanic rocks, the Sonoma Volcanics, that began

Plate 43.
Novato
Conglomerate
at Black
Point.

erupting about 9 million years ago to the east. These rocks are all more closely related to the geology of the North Bay than to that of Marin County and are considered in detail in chapter 10. Volcanic rocks make up much of Burdell Mountain, which is not a volcano. These volcanic rocks are resistant to erosion, and Burdell Mountain, at 1,558 feet, is the highest spot in northern Marin County. Burdell Mountain is surrounded by old landslides and slumps hundreds to thousands of years old. Rancho Olompali, a Native American village site on the northeastern side of Burdell Mountain, is located on large landslide deposits. The site has been dated at about 550 years before present, which gives us a minimum age of the landslide because the landslide must be older than the settlement on top of it.

As the Sonoma Volcanics erupted to the east, the sea encroached from the west and a shallow marine embayment covered much of southern Sonoma and northernmost Marin Counties. Deposits in this sea form the young marine sedimentary rocks that occur from the Sonoma county line west to Dillon Beach and Tomales. These sediments, the Wilson Grove Formation, are described in more detail in chapter 10.

Special Places to Explore

It would be difficult to find a more magnificent setting in which to explore the Bay Area's geology than southern Marin County. Here, three large areas of coherent Franciscan rock—Marin Headlands, Ring Mountain, and Mount Tamalpais—each surrounded by mélange, make up the geology underlying the landscape. Because all three are a park or preserve, their unique geologic treasures are available for everyone to explore. *A reminder: hammering on or collecting the rocks is not allowed in these special places.*

The Marin Headlands

The Marin Headlands, at the southern tip of Marin County, offers an outstanding view of San Francisco and the Golden Gate Bridge combined with excellent exposures of Franciscan rocks of the Marin Headlands Terrane (map 10). Although similar rocks are found elsewhere in the Coast Ranges, this is where they are best exposed. Therefore, this location has given its name to this particular combination of pillow basalt, radiolarian chert, and graywacke, wherever it is found.

The outcrops of chert in the headlands are truly spectacular. Here you can enjoy the beautiful color and form of these rocks (pls. 17, 44) as you think about their origin. Remember the uncountable numbers of radiolarian skeletons that make up each inch of this chert and the millions of years over which it accumulated (see chapter 3). All along Conzelman Road, west from Hwy. 101 at the north end of the Golden Gate Bridge, are exposures of tightly folded and contorted layers of chert and shale that represent over 100 million years of deposition, beginning about 200 million years ago. Although it is not easy to turn your back on the Golden Gate, stop at Battery Spencer and walk back down the hill for outstanding exposures of the chert.

Pillow basalt occurs further along Conzelman Road and at Kirby Cove, but the best ones make up the bold sea cliffs near the lighthouse at Point Bonita (pl. 45). Along the trail to the lighthouse you see large pillows piled up in layer after layer of lava flows. They look the same as those elsewhere in the Marin Headlands, but laboratory examination of these pillows shows that they have a different chemical composition and origin, so they are assigned to a different terrane, the Permanente Terrane.

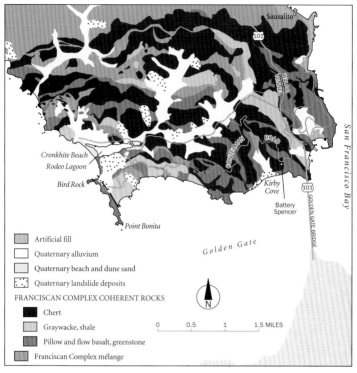

Map 10. Geologic map of the Marin Headlands.

LEGEND:

Artificial fill

Quaternary alluvium

Quaternary beach and dune sand

Quaternary landslide deposits

FRANCISCAN COMPLEX COHERENT ROCKS

Chert

Graywacke, shale

Pillow and flow basalt, greenstone

Franciscan Complex mélange

0   0.5   1   1.5 MILES

A drive north on Hwy. 101 from the Golden Gate Bridge takes you through the rock types of the Marin Headlands Terrane in the order in which they were formed long ago. If you are stuck in traffic, console yourself by looking as you go by. The north tower of the bridge is anchored in dark gray pillow basalt; next, red radiolarian chert (and some that has been bleached white) is exposed in road cuts on either side of the highway; and as you climb toward the Waldo Tunnel, the road is cut through gray brown graywacke and dark shale on the west (pl. 40) and bleached chert on the east. North of Waldo Tunnel there is a little more reddish chert just after the tunnel, then a long wall of dark gray pillow basalt.

Plate 44. Folded and faulted Franciscan radiolarian chert at Kirby Cove.

Plate 45. Point Bonita Lighthouse on pillow basalt.

Along the Headlands Shore

Ocean processes constantly form and change the landscape. The erosion-resistant Franciscan rocks of southern Marin County are under continuous attack by the sea, a battering that forms the high cliffs and sea stacks of the headlands. In summer, when ocean waves are gentle, it is hard to imagine how the ocean could have much effect. But visit the area in winter and you can feel the ground shake as huge waves pound the shore, shattering even the strongest rocks over time. Many feet of rock may tumble into the sea during a single heavy winter storm when the cliffs are repeatedly assaulted by huge waves and weakened by rain soaking into the ground.

In contrast, the quiet pocket beaches of the Marin Headlands are excellent places to enjoy the beauty of wave-polished Franciscan pebbles (pl. 18, *top*). Most of these little beaches are accessible only by a difficult scramble down the steep cliffs. Kirby Cove, however, is an exception. It is easily reached by walking down a gated road from Battery Spencer on Conzelman Road to a beach that overlooks the Golden Gate. Beautiful folded chert in the cliffs, sailboats, and freighters passing under the Golden Gate Bridge, which rises high over your head, make this a memorable walk.

Like most beach sands, those at Kirby Cove come from the surrounding area. Although light-colored mineral grains, such as quartz, are the most common grains in sands at Point Reyes Beach and Ocean Beach in San Francisco, here they make up only a small fraction of the sand. The headlands sand, like that at nearby Cronkhite Beach (pl. 7, C), is unusually dark because the source rocks are Franciscan. The reddish brown of chert grains is the dominant color, along with dark grays and browns of graywacke and pillow basalt. Scattered among these dark pebbles are the vibrant red, blue, green, and mustard yellow pebbles of altered chert, which tumble down from the ridge above Kirby Cove (pl. 18, *bottom*). A walk along the wet sand at the edge of the beach is a feast for the eyes, and the smooth wave-worn pebbles a pleasure to the touch. Franciscan rocks make some of the most beautiful beach sand in the world. But remember this is a national park, so leave the pebbles for others to enjoy.

Cronkhite Beach, at the end of the road past Rodeo Lagoon, is a popular destination for beach walkers, picnickers, and surfers (pl. 46). Because of its orientation to the ocean, this beach exhibits strong seasonal changes. In summer this is a quiet and peaceful landscape. The gentle summer waves bring sand up onto the beach but lack the energy to carry it back to the sea; thus the beach grows larger. Birds rest on Rodeo Lagoon, which becomes stagnant behind the mounded sand berm that cuts it off from the sea. But in winter, when storms pound the coast, the dramatic impact of ocean on land can be experienced here. As strong winter waves crash on the strand, retreating waves carry away beach sand and deposit it a few hundred feet from shore. The beach grows leaner, its angle steeper, and often at high tide, waves foam over the berm into the lagoon. When Rodeo Lagoon overflows, it cuts a channel through the berm to the ocean. At a very low tide in winter you can sit on Cronkhite Beach and watch the waves break over Potatopatch Shoal outside the Golden Gate (fig. 24). The shoal is composed of sand carried down from the Sierra Nevada during the Ice Age and deposited outside the Golden Gate (see chapter 6).

Plate 46. Cronkhite Beach.

Winter storms not only alter the shape of the beach but also bring a special treasure to the sand. Walk along Cronkhite Beach after a storm and scattered among the dark beach pebbles are tiny translucent reddish orange grains of carnelian, one of the silica minerals. No source of these little gems is exposed, and for a long time their origin was a puzzle to geologists. Then, as part of a project to document changes on the beach over a year's time, geologist Mary Hill discovered the source of the carnelian. During a particularly heavy storm in January 1967, the outflow from Rodeo Lagoon cut a channel almost 6 feet deep through the sand and exposed basalt bedrock. And there, filling tiny pockets in the rock, called vesicles, was carnelian. The vesicles are holes left by gases that bubbled out of the molten lava when it erupted beneath the ocean. After the lava cooled and hardened to rock, the vesicles filled with fluids rich in silica, and tiny carnelians formed, much like the agate in geodes.

The cliffs at the north end of the beach consist of Franciscan graywacke; those to the south are mainly pillow basalt and some chert. The individual pillows are not very distinct here because the basalt has been slightly metamorphosed. In this process new minerals have formed, giving the rock a characteristic greenish tinge; geologists call it greenstone (pl. 16). At a low tide this is an excellent place to examine such altered basalt with its veining of the white mineral calcite. Bird Rock and the other large sea stacks south of the beach are dominantly pillow basalt. They look white because of the bird guano coating them.

Ring Mountain: The Bay Area's Unusual Metamorphic Rocks

Ring Mountain on the Tiburon Peninsula is another of Marin's outstanding geologic locales, and it is also home to some very rare plant species, including several endemics, which grow nowhere else in the world. Because of its botanical and geologic importance, the site is a preserve. The geologic structure at Ring Mountain is quite straightforward—Franciscan graywacke and shale of the Alcatraz Terrane at its base and serpentinite on top. In between is mélange with remarkable high-grade Franciscan metamorphic rocks embedded in it (see chapter 3). Ring Mountain is one of the best locales in the Bay Area to see these metamorphics (pl. 47). Here you can wander among rocks in shades of blue and green, glittering with mica, covered with long dark blue or green crystals (the minerals glaucophane and hornblende) and sometimes speckled with tiny reddish garnets. These metamorphic rocks are mostly schist and eclogite, metamorphosed ocean crust basalt (pls. 21, 48). They are not only exquisitely beautiful but also geologically very significant. They were metamorphosed about 165 million years ago and probably represent the earliest stages of the Mesozoic subduction that produced the Franciscan rocks.

Above the mélange is serpentinite with some metamorphic blocks embedded in it. The serpentinite is characterized by a rubbly surface and a thin dark reddish soil. In spring it is covered

Plate 47. High-grade metamorphic block on Ring Mountain.

with wildflowers, including several rare species, such as the Oakland star tulip *(Calochortus umbellatus)*, and native grasses that are now uncommon in California, including purple needle grass *(Nassella pulchra)*. Few plants grow on serpentinite because the soil is poor in nutrients and contains elements that are toxic to many plants; however, several rare endemic plants, including the Tiburon mariposa lily *(Calochortus tiburonensis)*, serpentine reedgrass *(Calamagrostis ophitidis)*, Tiburon paintbrush *(Castilleja neglecta)*, and Tiburon buckwheat *(Eriogonum caninum)*, favor the unusual chemistry of serpentinite and grow among these rocks.

Mount Tamalpais: Cross Section through Subduction

Mount Tamalpais is a composite of different Franciscan rock types and terranes assembled during the Mesozoic plate collision. It has a very different geologic makeup than the Marin Headlands or Ring Mountain. The road to the top of Mount Tam (as it is known to most Bay Area residents) goes through the whole spectrum of Franciscan rocks, and at the top you are rewarded by a magnificent view of the northern Bay Area (pl. 30). You will find many good trails here from which to explore the interesting geology.

Plate 48. Close-up of mica schist.

The Franciscan rocks on Mount Tam include basalt, chert, graywacke, and shale. Most of the trails take you through extensive exposures of weathered graywacke, which also gives the soils their dominantly brownish color. Road cuts on the main road up to the peak expose basalt (between Pan Toll and Rock Springs) and reddish Franciscan chert in tightly folded layers similar to those found in the Marin Headlands. Much of the chert on Mount Tam has been bleached white, but you can recognize it by its typical layering. In other places the chert has been recrystallized to a bright red jasper. An extensive outcrop of serpentinite occurs at Rock Springs, an excellent place to get a feel for the rock, as well as the thin soils and rubbly surface characteristic of serpentinite outcrops.

It is not only Mount Tam's rocks that are geologically interesting; the mountain itself has a significant impact on the surrounding area by increasing the amount of rainfall in southern Marin County. Pacific storms rise over the mountain and drop much of their rain on its windward slopes, which record the highest annual rainfall in Marin County. Kentfield, on the north slope of the mountain, averages about 46 inches of rain annually, more than double the annual average for San Francisco (21 inches) or Oakland (18 inches). San Rafael, just 2 miles north of Kentfield, averages only 35 inches of rain annually. The high rainfall on the slopes of Mount Tam, combined with the mountain's geology, contributes to the large number of prehistoric landslides that have shaped it. The lower slopes on the north side are wrinkled and lumpy where landslides flowed down the mountainside long ago. Low spots in this landslide terrain filled with water to form small lakes, the predecessors of the present reservoirs — Phoenix, Bon Tempe, and Alpine Lakes and Lake Lagunitas.

Marin's many parks and trails provide fine access to some of the most interesting and varied geology in the Bay Area. Add a geologic map to your trail map, hiking staff, and bottle of water, and you can enjoy many hours in exploration of Marin's geology.

THE GEOLOGY HIDDEN BENEATH the city of San Francisco (map 11) tells a story of subduction, of rocks that have traveled thousands of miles to the Golden Gate, and of sand dunes that once stretched from Ocean Beach to the bay. The geology pokes so firmly through the cultural overlay (map 12) that with a little walking around one can learn a great deal, not only about the City's geology, but also about the Bay Area's early geologic history. The City's parks and shoreline, where the underlying Franciscan Complex bedrock is exposed in all its variety and beauty, provide excellent opportunities for geologic exploration.

Maps of San Francisco in gold rush days, when it was a small cluster of buildings around Yerba Buena Cove on San Francisco Bay, show rocky hills above a sea of sand. When we conduct our affairs in today's bustling city, it is not obvious that physically San Francisco is a small land mass. One can walk across it in a few hours; crest any high hill and see water—the bay, the Golden Gate, or the Pacific Ocean. A great deal of geology is concentrated in this small area.

One of the charms of San Francisco is its pattern of hills and valleys, which engage our senses while exercising our legs. The hills dominate the landscape: Telegraph, Nob, and Russian Hills in the east; the hills of the Presidio in the north; the central highlands, including Mount Sutro and Twin Peaks, which are San Francisco's highest; and San Bruno Mountain in San Mateo County to the south. This topography reflects the nature of the geology on which the City is built. As you might expect, strong and resistant rocks underlie the hills and the steep cliffs along the Golden Gate. The valleys are formed in softer, more easily eroded material. Much of the gently rolling terrain in the lowlands formerly was covered by extensive fields of sand dunes. Wetlands or shallow bays, now all filled, surrounded much of the City's northeastern perimeter (fig. 21). The financial district rests on filled-in Yerba Buena Cove, which lay between the steep bayshore cliffs of Telegraph Hill and Rincon Hill, the latter now the west anchorage of the Bay Bridge. A large embayment, Mission Bay, at the mouth of Mission Creek (fig. 21, A) was filled to create the South of Market area, and China Basin is the only water that remains there. San Francisco's picturesque North Beach District is also built on fill, as is the Marina District. This extensive fill, which has added more than 3 square miles to the City, now obscures its original margin. You can still locate some of the early shoreline of San

Francisco if you pay careful attention as you walk or drive its perimeter. Near the bay or Golden Gate where the land is quite flat, for example, along the Embarcadero or Bay Street, it is likely that you are on a surface underlain by fill. Then as you go toward the city center, a gentle rise of the street marks the former shoreline, and a steeper climb tells you where the hills began.

The Streams and Lakes That Once Were

Before the City was built up, many streams flowed from the highlands into lakes or to the bay and ocean. Only a few of them can be seen today; the rest are culverted and flow only underground. Lobos Creek, which still flows from Mountain Lake near the Presidio to the Golden Gate, is an exception. Juan Bautista de Anza camped along the shores of Mountain Lake in March 1776, when he first explored the San Francisco area. On that visit he also selected sites for a presidio (at Fort Point) and a mission, Mission Dolores, along Mission Creek (fig. 21, B). Lobos Creek provided San Francisco's first developed water supply, which was built by the San Francisco Water Works in 1858. Water flowed eastward from the creek through redwood flumes to the settlement at Yerba Buena Cove on the bay, the community that was to become the city of San Francisco. The young city used this water supply until 1900.

In central San Francisco a creek drained southward from Pacific Heights and Nob Hill toward what was the edge of the bay near what is now the intersection of 4th and Brannan Streets in the South of Market area (fig. 21, C). Another, Mission Creek, flowed from Twin Peaks along the present alignment of 18th Street to a small lake, Laguna de los Dolores, that existed near 17th and Mission Streets (fig. 21, D). It then flowed through the present South of Market area to an extensive marsh at the edge of the bay (fig. 21, A). When the 1776 de Anza expedition discovered the creek, they found a waterfall tumbling from the hills to the flatlands, probably at the change in slope at about 18th and Castro (fig. 21, E). Construction of Mission Dolores near this creek was started in 1782.

text continues on page 116

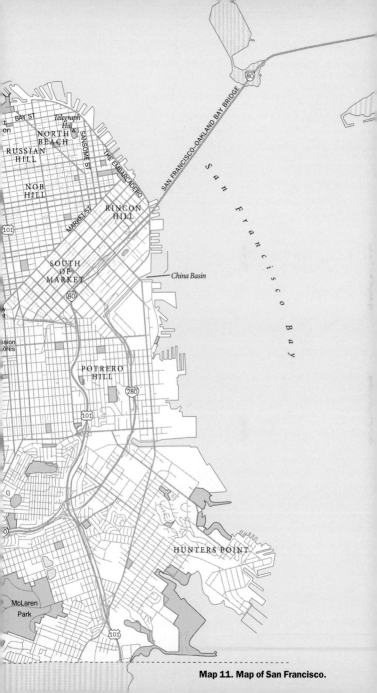

Map 11. Map of San Francisco.

Golden Gate

GOLDEN GATE BRIDGE

101

1

Golden Gate Park

1

19th AVE

1

35

35

Lake

Merced

35

1

280

SAN FRANCISCO
SAN MATEO

PACIFIC OCEAN

N

0   0.5   1   1.5 MILES

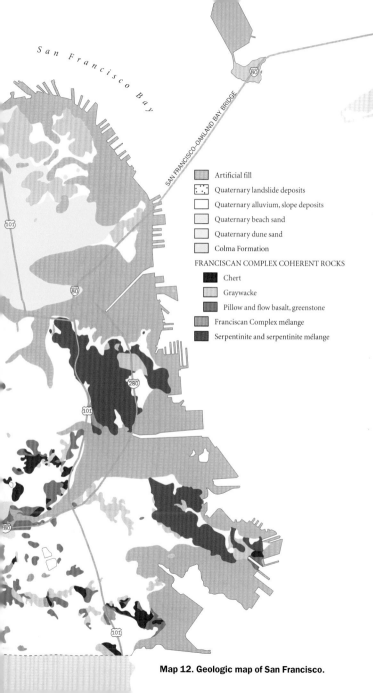

San Francisco Bay

SAN FRANCISCO-OAKLAND BAY BRIDGE

Artificial fill

Quaternary landslide deposits

Quaternary alluvium, slope deposits

Quaternary beach sand

Quaternary dune sand

Colma Formation

FRANCISCAN COMPLEX COHERENT ROCKS

Chert

Graywacke

Pillow and flow basalt, greenstone

Franciscan Complex mélange

Serpentinite and serpentinite mélange

Map 12. Geologic map of San Francisco.

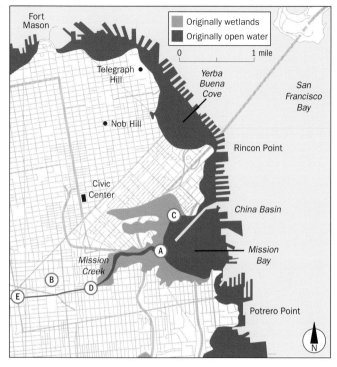

Figure 21. Extent of fill in San Francisco since 1847: A, mouth of Mission Creek; B, Mission Dolores; C, 4th and Brannan; D, Laguna de los Dolores; E, 18th Street and Castro.

At the western margin of San Francisco two prehistoric streams, one from the heights of Mount Davidson and the other from the hills of Daly City, flowed to the ocean before drifting sands cut them off and formed today's Lake Merced. The upper reaches of the streams were later filled, but you can still trace their courses on topographic maps of the City. At least 14 natural year-round lakes used to exist in what is today Golden Gate Park. The Chain of Lakes west of the buffalo paddock and Elk Glen Lake are remnants of these original lakes, but many others have been filled.

These filled shorelines, former creeks, and wetlands are among the most geologically unstable areas of San Francisco and consti-

Plate 49. House with former garages now below street level.

tute a major hazard during earthquakes. Unengineered fill settles unevenly over time and is prone to strong shaking during earthquakes. Areas of old fill were the most severely damaged areas in the City during the 1906 earthquake and were damaged again in the 1989 Loma Prieta earthquake. The old South of Market stream valley was filled with almost 200 feet of dune sand and garbage when the population exploded during the gold rush. Subsidence of the surface has been a problem in this area for a long time, especially in the lower parts of former Mission Creek, where the water table is close to the surface. If you walk around the area bounded by 5th and 6th Streets between Clara and Shipley, here and there you can still see the dramatic effect of decades of subsidence, although most of the older homes are gone. This area was destroyed by fire in the 1906 earthquake, and the oldest buildings date from just after the earthquake. The area subsided steadily after it was filled, and the streets and utilities were raised to grade in 1930 and 1950 to permit emergency vehicles better access. Some of the buildings were also raised to the new street level in 1930, but the cost was too high in 1950, and the buildings were not realigned again. Today you can still find a few older buildings with front entrances or garages as much as 6 feet below present street level and new entrances cut to the second floor (pl. 49). Many of the older buildings show the effects of uneven settling, where one part of a building has settled more than another. Cracked and tilted buildings are the result.

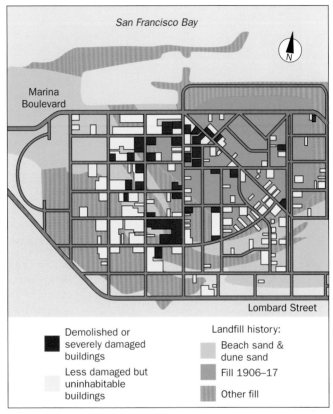

Figure 22. Pattern of damage in Marina District from the 1989 Loma Prieta earthquake.

Many new buildings were built in the South of Market area during the boom of the 1990s. They require special engineering to withstand the continuing subsidence in this area. One owner built a steel frame house that can be realigned if it settles unevenly. During the Loma Prieta earthquake the building was undamaged but one wall settled several inches. The owner jacked the side of the building up, realigned the structure, and had a level house again 4 days after the earthquake.

The Marina District, much of which is built on lowlands filled in 1913 for the Panama-Pacific Exposition, also experienced severe shaking in the Loma Prieta earthquake (fig. 22). The ground settled as much as 5 inches, and many homes were destroyed by shaking and fire. The damage was concentrated in areas where fill had been placed on top of the original beach and dune sand. Ironically, some of the rubble used as fill in 1913 was debris from the 1906 earthquake.

The City on the Dunes

More than half of San Francisco, from Ocean Beach to the bay, was once covered by sand dunes (map 12, fig. 23). An 1857 geological report noted that the area of dunes "has the aspect and character of a desert." Sand such as that which today regularly blows across the Great Highway along Ocean Beach once lay more than 100 feet deep east of Golden Gate Park. Some dunes were more than 60 feet high. Strong winds blew sand over San

Figure 23. The sand dunes of western San Francisco, looking southwest across Golden Gate Park, 1924.

Francisco's hills (sand once covered the 575-foot summit of Buena Vista Park) and caused sand to drift as far east as the bay. Only the wind-sheltered areas east of Mount Sutro and Twin Peaks were free of sand. Even today, construction uncovers the extensive sand fields. Residents of the Sunset District are mostly unaware that their lawns and flowerbeds are rooted in imported topsoil covering dune sand. But if you dig a hole deep enough you can find it. Remnants of dunes can be seen in a few scattered locations around the City without digging a deep hole in your backyard, such as along the west side of Mount Sutro and at the south end of Bakers Beach.

Ice Age Dune Sand

As remarkable as dunes under San Francisco may seem, the story of the origin of the sand is even stranger. Some of this sand comes, as expected, from local rocks such as the sandstone cliffs near Fort Funston. A reverse eddy south of the Golden Gate carries it northward to Ocean Beach. Much of the sand has a surprising source—the Sierra Nevada 150 miles away. During the last ice age, which was at a maximum 20,000 years ago, the granitic rocks of the Sierra Nevada were ground down by glaciers, and the resulting sediment was carried off by great rivers draining the mountains. Larger rocks were deposited in Sierra streams, but the small sand grains traveled westward, carried by tributaries to the Sacramento and San Joaquin Rivers, then across the Central Valley, through Carquinez Strait, and out to the ocean. Unimaginable volumes of rock debris were moved from mountain to sea in this process of making high places low.

The Ice Age sand was deposited along river banks from the Sierra Nevada to the coast. Because sea level was about 400 feet lower during the last glaciation than it is today, the shoreline lay far to the west of the present shore, out beyond the Farallon Islands, and San Francisco Bay was a wide river valley (see chapter 6). The rivers flowed through the valley and the Golden Gate, and westward across the then-exposed continental shelf. The sand carried by these rivers was deposited along the distant shore during annual floods. It was then picked up by the prevailing westerly winds and blown back across the exposed continental shelf to cover San Francisco and parts of the central bay as far east as Alameda and Oakland.

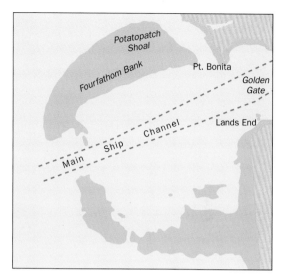

Figure 24. The San Francisco bar (shaded blue semicircle) outside Golden Gate. Potatopatch Shoal and Fourfathom Bank are part of the bar.

The San Francisco bar, a large crescent-shaped deposit of sand that lies outside the Golden Gate (fig. 24), is composed of local and Sierra sand. Today a channel must be dredged through the bar to permit large ships to enter San Francisco Bay. At very low tides you can see waves breaking over the Potatopatch Shoal, the northern part of the bar. This name apparently dates from the late 1800s, when potatoes were grown on farms around Bolinas Lagoon in Marin County and were shipped to markets in San Francisco. Occasionally a potato boat would capsize on the sand bar, spilling its load of potatoes.

From Dunes to Arboretum in Golden Gate Park

Looking at the rolling lawns and rhododendron groves of Golden Gate Park today, you would never guess that beneath the greenery lie Ice Age sand dunes. To make a park out of more than 13,000 acres of sand dunes was not an easy task. Removing such a vast quantity of sand was clearly impossible. Early park workers studied old European techniques for sand reclamation and laboriously, foot by foot, planted vegetation to stabilize the sand. First they planted barley and lupine seeds; the quick-sprouting barley provided protection for the slower-growing lupine. Because lupine could survive in the sand for several years, it provided a good cover for the small trees that were planted next. Near the ocean, where the sand was too salty for barley and lupine, beach grasses were planted to stabilize the dunes. In just a few years, from 1872 to 1875, much of the dune sand was covered with lupine and young trees. Roads and paths were installed, and Golden Gate Park was born.

Modern Sand along the Shore

Today, sand still accumulates along Ocean Beach (pl. 50) and blows inland toward the City. Cement walls vainly try to keep the drifting sand in check, but City work crews regularly close the Great Highway to return sand to the beach. This sand, which comes in part from local sandstone and in part from the older Sierra Nevada sand, has a mineral composition similar to that of the sedimentary rocks exposed in the cliffs at Fort Funston. These weakly cemented sandstones, called the Merced and Colma Formations (pls. 84, 86), are easily fragmented by ocean waves and intense rainfall. Landslides, which are common along this shore, contribute large amounts of sand to the beach. The older Sierra sand from the continental shelf is mixed with sea cliff sand to form the unusual combination of grains found on the beach.

The next time you go to Ocean Beach, take along a magnifying glass or hand lens and a magnet for a closer look at the sand (pl. 7, E). You can see that it consists of a great variety of grains of different-colored minerals and tiny rock fragments polished smooth and rounded by constant rolling back and forth in the waves. The grains consist mostly of colorless clear quartz, opaque milky feldspar, fragments of reddish chert and greenish serpentine, black magnetite (which is magnetic), and occasional pink garnets. Because magnetite is denser and heavier than the other minerals, it gets concentrated into delicate drifts of black sand on the surface of the beach as the waves wash back and forth. This sorting occurs when gentle waves have just enough energy to pick up and carry away the lighter grains, but not enough to move the

Plate 50. Ocean Beach, looking south from the Cliff House at Point Lobos. Rocks on left are graywacke of the San Bruno Mountain Terrane.

heavier grains. Also among the heavy grains are minute quantities of gold, which at times have yielded a few ounces to very patient gold panners.

Under the City: The Rocky Foundation

The rocks exposed in San Francisco's hills and buried beneath the City's urban landscape and surface deposits are the interesting Franciscan Complex rocks so common in the Coast Ranges (see chapter 3). Brush away the buildings and streets, and you would see a colorful geologic palette of the typical Franciscan rocks—dark gray pillow basalt, red chert, gray brown graywacke sandstone. Graywacke is the most common—the entire northeast and southwest parts of the city are underlain by it. The other Franciscan rock types and serpentinite are exposed in a band across the middle of the City.

Such variety exists because the bedrock beneath the City consists of three different Franciscan terranes (map 13), each with its own characteristic rock types (table 2). The terranes cut across

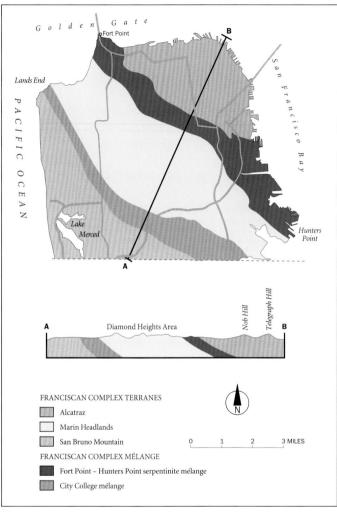

Map 13. *Top,* Franciscan terranes: San Bruno Mountain Terrane, City College mélange, Marin Headlands Terrane, Fort Point–Hunters Point serpentinite mélange, Alcatraz Terrane. *Bottom,* Cross section from A to B showing interpretation of structure below the surface.

San Francisco in a northwest-southeast direction, reflecting the order in which they were accreted to the continental margin. The Alcatraz Terrane on the northeast was the first accreted; the Marin Headlands Terrane was next; the San Bruno Mountain Terrane was the last. These terranes are separated by two mélanges, the Fort Point–Hunters Point mélange between the Alcatraz and Marin Headlands Terranes; and the City College mélange between the Marin Headlands and San Bruno Mountain Terranes. The bedrock geology of San Francisco is best described in terms of these terranes.

Alcatraz Terrane

The north and east parts of the City are underlain by the Alcatraz Terrane, which consists mostly of graywacke. The terrane is named after the rock found on Alcatraz , as you might guess, and it is also found on Yerba Buena Island, the natural island that anchors the middle of the Bay Bridge. Fort Mason rests on it, as does the eastern part of the Presidio. One of the best exposures is at Telegraph Hill, where graywacke can be seen in a steep quarry face at the foot of Lombard Street (pl. 19) and along Sansome Street. Both Russian and Nob Hills are also graywacke. Like other graywackes, this rock was formed by turbidity flows (see chapter 3). Because the Telegraph Hill graywacke is massive—that is, without layers—it was probably deposited by one or two very large flows, rather than by many smaller ones. The graywacke is highly fractured, which gives it an appearance of layering, but these fractures formed after the original sediments had hardened into rock, probably as the Alcatraz Terrane was accreted to the North American continent or faulted to its present position. Fossils found on Alcatraz indicate that this graywacke is 130 to 135 million years old.

The Telegraph Hill graywacke has been quarried since the early days of San Francisco. During the gold rush the rock was used as ballast for sailing ships that brought supplies to the new city and were returning to their home ports empty. In this way Telegraph Hill graywacke was carried around the world. In the 1850s and 1860s it was quarried to create sea walls for the harbor and for fill to build the Embarcadero. The steepened slopes of the quarry are prone to landsliding, and numerous rockfalls have occurred, including a large one in the 1970s, during which the upper end of Lombard Street failed and its granitic curbstones fell over the cliff. Now they lie incongruously among the graywacke rubble at the foot of the cliff.

Marin Headlands Terrane

The rocks in central San Francisco are the most colorful in the City; they belong to the Marin Headlands Terrane and underlie the highest hills of the City. The rich variety of Franciscan rocks in this terrane can be appreciated in excellent outcrops, such as at Twin Peaks, Mount Sutro, Glen Canyon Park, Corona Heights, and the eastern half of Golden Gate Park. The two summits of Twin Peaks consist of radiolarian chert lying on top of ocean crust basalt. Both are hard rocks that resist erosion, which is why Twin Peaks are such high hills (the elevation of the south peak is 922 feet). On the west side of the north peak you can see tightly folded chert and basalt. They appear to be in contact with much less resistant serpentinite mélange on the east side, where the hill drops off steeply.

Other fine exposures of radiolarian chert occur along O'Shaughnessy Blvd., on the west side of Glen Canyon Rock, at Rock Outcrop Park in Golden Gate Heights, and at Corona Heights, where beautifully folded chert is exposed behind the Randall Museum (pl. 51). At the base of the hill on the north side of Corona Heights, at Peixotto Playground (built in a former quarry), is an impressive example of chert with slickensides (pl. 52). Slickensides are smoothly polished and grooved surfaces formed by the grinding of rock against rock under intense pressure. They are common in soft serpentinite; this is one of the few places where they occur in the much harder chert. Although the slickensided chert looks water polished or even wet in a certain light, the slickensides are formed tectonically, not by water.

Graywacke of the Marin Headlands Terrane is exposed at Lands End (pl. 53) and in Lincoln Park, east of Lands End, where one of the rare large Franciscan fossils was found. It was an ammonite, ancestor of today's chambered nautilus, and it confirmed the age of these Franciscan rocks as Cretaceous.

The Golden Gate separates the Marin Headlands Terrane in San Francisco from the much larger exposures of this terrane in the Marin Headlands to the north (see chapter 4). In both locations the rocks are similar, but because the terranes are oriented

Plate 51. Prominent folds in chert at Randall Museum, Corona Heights.

Plate 52. Author at chert slickensides at Peixotto Playground.

Plate 53. Graywacke of Marin Headlands Terrane at Lands End.

differently on either side of the Golden Gate, there is probably a fault between them. Such a fault would also help explain the Golden Gate itself, because faults are zones of weakness where rock is easily eroded to form valleys.

San Bruno Mountain Terrane

The westernmost terrane in San Francisco is the San Bruno Mountain Terrane, which extends from the sea cliffs at Lands End and Point Lobos southeast beneath the former dunes to Daly City. Most of the long, high ridge that forms San Bruno Mountain lies in Daly City, and its geology is discussed further in chapter 7. The Lands End area has some of the most spectacular scenery in San Francisco. From Point Lobos on a clear day you can see along the coast from Point Reyes in the north to Point San Pedro and Montara Mountain to the south.

The San Bruno Mountain Terrane is composed of Franciscan graywacke and shale. Like the other Franciscan graywackes, the rocks are formed of sediments deposited by turbidity flows. Although this terrane is very similar to the Alcatraz Terrane in rock type, it is younger and different in mineral composition and in appearance. It consists of both massive graywacke and layers of sandstone and shale deposited by many smaller turbidity flows (pl. 54). The graywacke forms the cliffs and sea stacks at Point

Plate 54. Graywacke and shale in turbidites of San Bruno Mountain Terrane at the former Castle Lanes Bowling Alley, now a condominium complex and not accessible. Photo taken 1988.

Lobos and underlies the Cliff House and Sutro Baths foundation there (pl. 55); it is highly fractured and prone to landslides. Near the Cliff House the graywacke has been covered with concrete to protect the highway from falling rock. Historically, the many sea stacks just offshore, which are erosional remnants of resistant graywacke, were a considerable hazard to navigation. At low tide you can still see the remains of several ships that ran aground on these rocks in days before bar pilots guided ships through the Golden Gate into San Francisco Bay.

Mélange Zones

The valleys of central San Francisco are largely underlain by two mélange zones, the City College mélange and the Fort Point–Hunters Point serpentinite mélange. These mélanges were probably formed as the Franciscan rocks were scraped off the sinking plate during subduction and were plastered onto the North American Plate. The ground-up serpentinite or graywacke that forms the mélange matrix is soft and easily eroded. Before the City was paved over, creeks carried away the soft rock, wearing the landscape down to its present topography. Here and there

Plate 55. Graywacke of San Bruno Mountain Terrane at Point Lobos with foundations of former Sutro Baths. Note brown graywacke cliffs and gray shale slope.

in both mélanges, large blocks of more resistant rocks form low hills. The resistant blocks in the City College mélange are basalt, coherent serpentinite, graywacke, and metamorphic rocks. The Fort Point–Hunters Point mélange has huge resistant blocks that are largely serpentinite. They include Potrero Hill, Fort Point, and the hill at Market and Duboce Streets on which the new mint sits. Both Fort Point and the south anchorage of the Golden Gate Bridge are built on serpentinite. The fort is an excellent place to learn about the early history of San Francisco and to get a good look at our state rock. Resistant blocks of serpentinite in a range of shades from pale greenish blue to almost black are exposed along the sea cliffs near the fort and at Bakers Beach to the southwest (see chapter 2, opening photo). The structure of the serpentinite mélange is also well exposed near the new mint, where resistant knobs of serpentinite are embedded in a crushed serpentinite matrix.

On Top of the Basement Rocks

Before the City was developed, the sand dunes described earlier covered the bedrock under the lowlands of the northern half of San Francisco, and young sedimentary deposits of the Colma

Formation covered lowlands in the southern half (map 12). The Colma Formation underlies most of the southwestern part of the City from Lake Merced to McLaren Park. It is a yellowish brown, soft, unconsolidated sand that has not been cemented into solid rock. This formation was deposited in coastal and estuarine environments during the last interglacial period, about 80,000 to 125,000 years ago, when sea level was about 20 feet higher than today. Here it includes dune and beach sands; on the Peninsula much of the Colma was deposited along a shallow seaway between the ocean and San Francisco Bay. In San Francisco, most of the Colma Formation has been eroded away or is covered with houses and streets. The best place to see it is in the sea cliffs south of Ocean Beach (see chapter 7).

In southeastern San Francisco, the bedrock is overlain by sediments eroded out of the central San Francisco hills. Much of this eroded rock has piled up at the base of the hills and fills the valleys between them. Some has been transported by creeks that flowed to the bay. Because the hills are composed of Franciscan rocks, the overlying sediments include debris from the different types of Franciscan. You can find reddish chert rubble and brownish graywacke deposits east of Twin Peaks and McLaren Park and serpentinite sediments along the base of Potrero Hill. Most of this area has been built up, but here and there you may find some young sedimentary material and can most likely recognize the source of the sediment by the color of the deposit.

By exploring San Francisco's sea cliffs and hills in its many wonderful parks, you can discover that there is more geology visible in the City than you might guess. You can become acquainted with the Franciscan rocks that underlie the City and that tell the story of almost 200 million years of geologic activity. You can touch and feel our state rock, serpentinite. You can see remnants of young sand dunes that once covered most of the area. And best of all, you can make most of these explorations by foot and public transportation, as geologist Clyde Wahrhaftig did in his classic 1984 booklet, *Streetcar to Subduction* (see the Further Reading section at the end of the book). Although a few of the places he described are no longer accessible, in most cases nearby exposures show the local geology, and the sites in parks and other public lands are still there to be enjoyed by all.

SAN FRANCISCO BAY is the jewel in the heart of the Bay Area, a delight to the eye for the millions who live along its shore, and essential to the Bay Area's economic life. The bay (map 14) is also full of geological surprises.

The bay is connected to the Pacific Ocean through the Golden Gate, so named by an early California explorer, General John C. Fremont, in reference to Chrysoceras, the harbor of Byzantium. Although the California coast was sailed by many explorers, traders, and pirates after the first explorations in the sixteenth century, San Francisco Bay itself was not discovered by Europeans until almost three centuries later because the Golden Gate is both narrow and commonly fog shrouded during the summer. In 1769 a party led by Don Gaspar de Portolá approached by land and was the first to sight the bay (pl. 56). Six years later the first Europeans sailed through the Golden Gate.

The present San Francisco Bay is surprisingly young, only about 10,000 years old, a product of today's high sea level. The bay is also ephemeral, coming and going as sea level changes. The secrets of its past are buried in the mud beneath the shallow waters. They tell us that the valley in which the bay lies was first flooded with ocean waters perhaps a million years ago, and that bay waters have come and gone several times. The valley itself is tectonic, a downdropped block between the San Andreas and Hayward Faults.

Plate 56. Plaque on Sweeney Ridge east of Pacifica commemorating the first sighting of San Francisco Bay by Gaspar de Portolá's expedition on November 4, 1769.

The Estuary and Its Surprises

San Francisco Bay is an estuary, an embayment where freshwaters and saltwaters mix. Here, the freshwaters of the Sacramento and San Joaquin Rivers meet the marine waters that pour through the Golden Gate. Scientists now commonly refer to the "San Francisco Estuary," rather than "San Francisco Bay," to make it clear that San Pablo Bay, Carquinez Strait, Suisun Bay, and the Delta are all part of the estuarine system.

An astounding amount of freshwater flows through the bay. The Sacramento and San Joaquin Rivers together carry the runoff from 40 percent of California. The watershed of the estuary extends from the Cascade Range and Klamath Mountains in the north to the Tehachapi Mountains in the south, and from the Sierra Nevada crest on the east to the crests of the Coast Ranges on the west (fig. 25). Approximately 90 percent of the freshwater that flows into San Francisco Bay comes through the Delta, primarily during the winter rainy season and from spring snowmelt in the mountains. The remaining 10 percent comes from local streams, such as the Petaluma and Napa Rivers, and from wastewater treatment plants. Today, the inflow is only about 60 percent of the historic flow before 1850 because freshwater is diverted for municipal needs in the Bay Area and southern California (15 percent) and for agricultural uses in the Central Valley (85 percent).

That the bay is only 10,000 years old is not its only surprise; it is also much shallower than it looks (fig. 26). As you cross it on one of the bridges, you would not guess that the depth of the bay averages about 18 feet. With a few exceptions, the waters are deep only in parts of the central bay between San Francisco, Marin, and Angel Island; in the dredged channels that provide access to ports in Oakland, Richmond, and Redwood City; and in natural channels through Carquinez Strait (about 90 feet deep), Raccoon Strait (60 feet), and the Golden Gate, much of which is between 100 and 200 feet deep. Water depth in the bay changes twice a day with the tides. San Francisco Bay tides are measured by a tide gauge at the Presidio. It is the oldest continuously operating tide gauge in the Western Hemisphere, established in 1854. San Francisco Bay has two high and two low tides of unequal height each tidal day. We refer to these tides as mixed (of unequal height) and

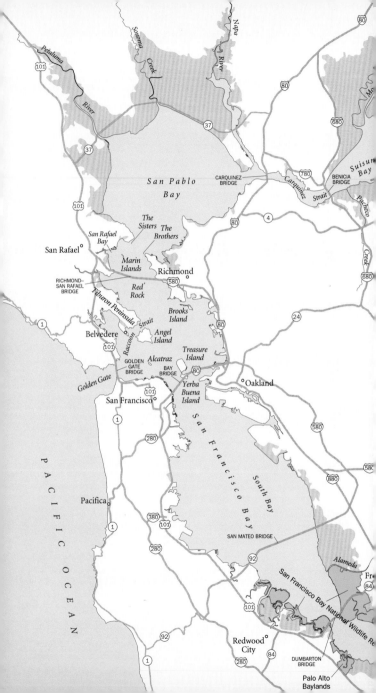

Map 14. The San Francisco estuary, comprising San Francisco, San Pablo, and Suisun Bays. The shaded blue represents intertidal areas between the highest and lowest tides.

Plate 57. Palo Alto Baylands marsh at low tide, showing tidal channel and mudflat.

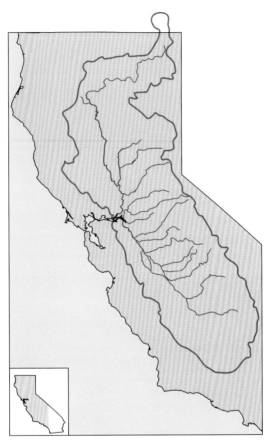

Figure 25. Watershed of the San Francisco Estuary.

semidiurnal (occurring twice a day) (fig. 27). The average tidal range, which is the difference between consecutive high and low tides, is about 4 feet in the central bay, 5 feet in northern San Pablo Bay, and 7 feet at the south end of the bay. Tides and river inflow combine to move very large volumes of water in and out of the bay daily. The volume of water that enters and leaves the bay with each tide, the tidal prism, is about one-quarter of the bay's volume.

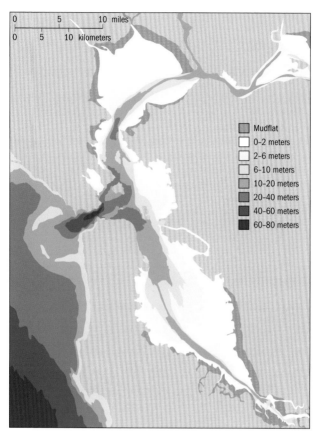

Mudflat
0–2 meters
2–6 meters
6–10 meters
10–20 meters
20–40 meters
40–60 meters
60–80 meters

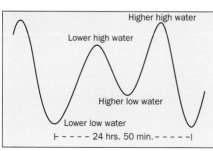

Higher high water

Lower high water

Higher low water

Lower low water

⊢ – – – – 24 hrs. 50 min. – – – – ⊣

Figure 26 *(above)*.
Bathymetry of
San Francisco
Bay. 1 meter
= 3.3 feet.

Figure 27 *(left)*.
Tidal cycle.

Water That Flows "Upstream"

Hidden under the bay's shallow water are still other surprises. The water you usually see at the surface of San Francisco Bay is the fresh and warm water discharged by the Sacramento and San Joaquin Rivers into the estuary. But below the surface near the bottom of the bay, a flow of cold, salty, and dense marine water moves landward beneath the freshwater (fig. 28). This pattern of two-layer circulation is typical of many estuaries. Under present water conditions, the marine and fresh bottom waters usually meet and mix toward the eastern end of San Pablo Bay. But during dry summers and periods of drought, salty marine bottom waters may flow upstream as far as the Delta. When this salt-water wedge moves far upstream, it adversely affects industrial, municipal, and agricultural water users that withdraw water from the upper end of the estuary. Conversely, during very wet years or during the seasonal high runoff from winter storms, you can sometimes see a prominent line where brown, sediment-laden freshwater flows over the denser ocean water (pl. 58).

In the upstream part of the estuary where freshwater enters the system, average salinity in winter (the wet season) is low, ranging from 1 part per thousand at the Delta to about 20 parts per thousand in San Pablo Bay. Compare this with the average marine salinity of 35 parts per thousand. At the Golden Gate and in the South Bay winter salinity averages about 25 parts per thousand. Summer salinities are higher because less freshwater flows

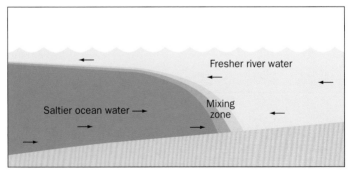

Figure 28. Two-layer circulation of San Francisco Bay.

Plate 58. Layer of muddy freshwater over saltwater near Point Bonita, March 1958. Golden Gate Bridge is in background.

into the bay, and the average is about 30 parts per thousand in the South Bay and about 33 parts per thousand at the Golden Gate. As a result of the geography of the bay, circulation is poor in the South Bay, which receives little natural freshwater inflow. Today about 90 percent of the freshwater inflow to the South Bay comes from wastewater treatment plants.

Secrets of the Sediments

Several hundred feet of soft mud and sand lie beneath the waters of the bay. For hundreds of thousands of years, sediment has been eroded from the watershed and carried to the bay by the rivers, and bits of California's mountains have come to rest beneath its waters. When sediment-laden freshwater meets the bay's saltier water, clay-size particles flocculate (clump together) and, along with coarser grains, sink to the bottom. Today, most of the sediment from natural erosion in the mountains is trapped behind dams on Sierra streams, and runoff from agricultural lands provides much of the mud deposited in the bay. A large amount of it initially settles in Suisun and San Pablo Bays. Spring and summer winds then create waves that resuspend this sedi-

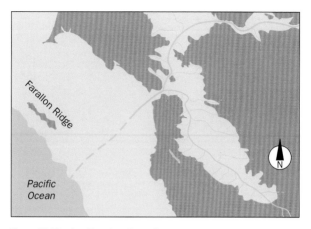

Figure 29. The San Francisco Bay valley and continental shelf at peak of the last ice age, when the shoreline was west of the Farallon Islands. Gold is present land; beige is area that is water today but was land 20,000 years ago. River dashed where approximate.

ment, and tidal currents distribute it throughout the bay system or carry it out the Golden Gate.

During the period of hydraulic gold mining in the Sierra Nevada, vast quantities of sediment were washed into the Sacramento River and brought to San Francisco Bay. In 1917 G. K. Gilbert, a noted geologist, calculated that more than 1 billion cubic yards of mining debris was deposited in the northern part of the San Francisco estuary between 1849 and 1914. Large areas of Suisun and San Pablo Bays were shoaled by this sediment, which is still being redistributed today by tidal currents.

The history of San Francisco Bay is recorded in the sediments that have accumulated over the past million years or so. We can read this record by drilling deep into the bay and bringing up core samples of the mud. Many such cores have been drilled for bridge foundation studies, and the sediment in them tells us that bay waters have come and gone several times. During interglacial times, such as the one in which we are living today, sea level is high; when glacial ice builds up on land, sea level drops. In the past, sea level has risen and fallen many times as ice ages waxed and waned.

The upper several hundred feet or so of sediment in the bay

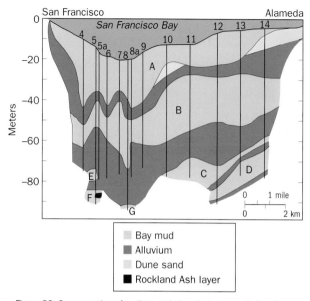

Figure 30. Cross section of sediments in boreholes beneath San Francisco Bay, showing alternating estuarine (bay mud) and terrestrial (alluvium) sediments. Layer A is the sediment accumulating beneath San Francisco Bay today. Layer B was deposited during the last interglacial period. Estuarine layers E, F, and G may correlate with C or D. A layer of ash in F at the base of borehole 5 has been dated at approximately 570,000 years. Vertical exaggeration is 100 times.

have accumulated in the present estuary in the past 10,000 years. Below them lies an interval of alluvial sediment and dune sand. Sea level must have been low and the valley must have been dry land when the sediment and sand were deposited. Can you imagine what the Bay Area would look like without the bay? Take your time machine back 20,000 years. Below you is a broad river valley (fig. 29), its grass golden in summer, green in winter. Go back even farther, to the last interglacial period 120,000 years ago, for example, and you see water again. Sea level must have been high. Sediment cores taken in the South Bay contain deposits of at least four separate estuarine units with alluvial (terrestrial) deposits between them (fig. 30), reflecting changes in sea level over about the last half a million years.

During the last interglacial period, the bay was a good deal larger than it is today because sea level was about 20 feet higher. Evidence from microscopic organisms preserved in the mud beneath the bay tell us that conditions in the bay were similar to those you find along the coast of Bay Area counties today. The organisms provide evidence that the last interglacial bay was also saltier, colder, and deeper.

Although no glacial ice ever covered the Bay Area, glaciation elsewhere profoundly affected our landscape during the ice ages. As ice repeatedly built up on land and sea level dropped, shore-lines shifted oceanward. During the peak of the Ice Age, about 20,000 years ago, sea level was about 400 feet below its present level. At that time the shoreline was just beyond the Farallon Islands, 19 miles west of San Francisco (fig. 29). On a sunny weekend you could have hiked out across the broad, gently sloping continental shelf for a picnic on Farallon Ridge overlooking the

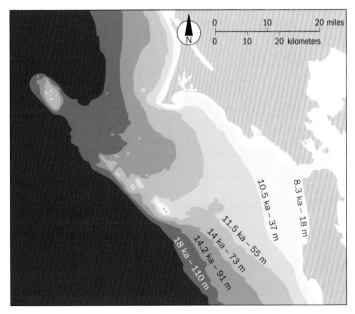

Figure 31. Rising sea level as glaciers melted after the last ice age. Time in thousands of years ago (ka); depth in meters.

Pacific Ocean to the west. Or you could have stood on the headlands above the Golden Gate and watched the mighty flow of the river as it poured through the valley, tumbling over great cascades or waterfalls. A roar would have filled the air as the meltwater from far-off glaciers sent floods through the Golden Gate to the ocean. The river cut deep channels into the bedrock at Carquinez and Raccoon Straits, turning the tip of Tiburon Peninsula into an island that became Angel Island. The river also carved the canyon through the Golden Gate, and strong tidal currents scoured out a depression 364 feet deep just west of the Golden Gate Bridge. About 18,000 years ago, as the glaciers began to melt, sea level rose again. The rising sea reentered the Golden Gate about 9,000 years ago, gradually filling the valley and creating today's bay (fig. 31).

Early History of the Bay

The Sacramento–San Joaquin River sysem has not always flowed through the Bay Area. Until about 3 million years ago, rivers that drained to the Central Valley flowed southward to an ocean embayment that existed along the west side of the San Joaquin Valley in the Kettleman Hills. As the valley gradually emerged from the sea and movement along the San Andreas Fault closed off that outlet, a large lake formed in the Central Valley. It expanded during glacial times, when the climate was cooler and wetter, then shrank during warmer times between ice ages. About 620,000 years ago near the end of an unusually long and intense glacial period, the lake spilled over a low area in the Coast Ranges, carved Carquinez Strait, and began to flow through the Bay Area. Evidence for this dramatic event is present in sediments in the sea cliffs south of San Francisco, where there is an abrupt change in minerals from those derived from local rocks to minerals from Sierra Nevada rocks.

A layer of volcanic ash, the Rockland Ash (pl. 85), is present above the mineral change in the sea cliffs and also occurs in muddy sediments almost 400 feet beneath the bay (fig. 30). The presence of the Rockland Ash in the bay mud shows that by the time the source volcano near Lassen Peak erupted, about 570,000 years ago, the Central Valley was draining through the bay. The opening to the ocean through the Golden Gate may not have existed at that time. Sediments on the Peninsula suggest that the

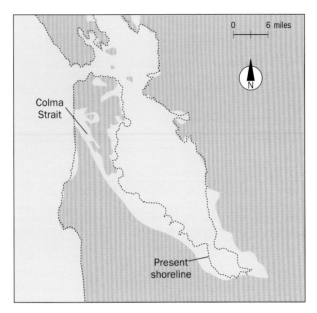

Figure 32. Drainage through Colma Strait during last interglacial period, 125,000 years ago. In this sketch San Francisco is shown as an island, but the Golden Gate may not have been open at this time, and San Francisco may have been connected to Marin County.

river drainage may have first flowed to the ocean through the "Colma Strait," a narrow seaway that crossed the Peninsula south of San Francisco (fig. 32). The change in drainage from the Colma Strait to the Golden Gate probably occurred sometime in the past several hundred thousand years.

Marshes and Plains around the Bay

Much of the urban population of the Bay Area lives and works on the gently sloping plain between the hills and the water. The necklace of lowlands that circles San Francisco Bay was formed by two geologic processes, one that uplifted the hills and the other that wore them down. Both were going on at the same time. As the hills around the bay were eroded, streams carried bits and pieces of rock downhill and deposited them at the edge of the bay. During the rainy season, and particularly in large storms, creeks brought much material out of the hills. Sediment built up where the creeks lost energy and dropped their load. Over thousands of years these sediments formed alluvial fans at the base of the hills. Eventually, the individual alluvial fans joined to form the alluvial plain on which many of us live and work.

The sediments of the alluvial plain consist of coarse sand and gravel deposited during storms and finer sand and silt deposited during quiet flow. These successive layers build up the plain. The coarser layers hold water and are major aquifers, an important source of water for communities around the bay. Over the past hundred years, however, so much freshwater has been pumped out of the aquifers that the water table has fallen and saltwater has seeped into the aquifers from the bay. Saltwater intrusion has been a particular problem in the Fremont area in southern Alameda County, where municipal wells became polluted in the middle of the twentieth century. Today Alameda Creek waters are diverted into former quarries at Quarry Lakes Regional Recreation Area in order to recharge the aquifers with freshwater.

The shoreline of the present bay is mostly artificial fill placed over the past 150 years. As a result, the present surface area of the bay is almost 40 percent less than its historic surface area. Before the margins of the bay were filled, they were ringed by extensive mudflats and marshes (fig. 33a). These wetlands were home to countless invertebrates that were food for small fish and birds; they were the nursery for most of the organisms that lived in the bay. Now, about 95 percent of those wetlands have been destroyed by diking and filling (fig. 33b)—for agriculture and industry, for houses, airports, and shopping centers. And for more than 100 years cities around the bay dumped their garbage at the water's edge. Few parks existed along the bay, and the public had access to only a few miles of the shore. In the first half of the twentieth century the marshes and mudflats were infamous for odor from the raw sewage that cities discharged into the bay.

Since the 1960s a strong movement to stop the loss of wetlands and restore access to the bay has reversed this trend, and raw sewage no longer flows into the bay. Now healthy mudflats and marshes can be found at many places. Soon a 400-mile bay trail

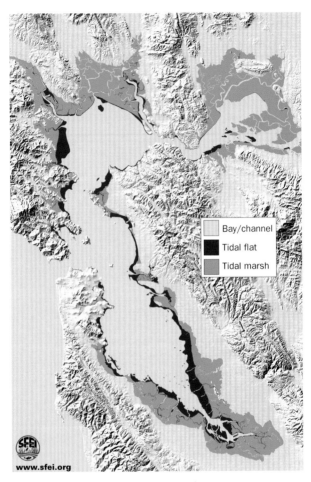

Figure 33a. Historic wetlands of San Francisco Bay.

Bay/channel
Tidal flat
Tidal marsh

www.sfei.org

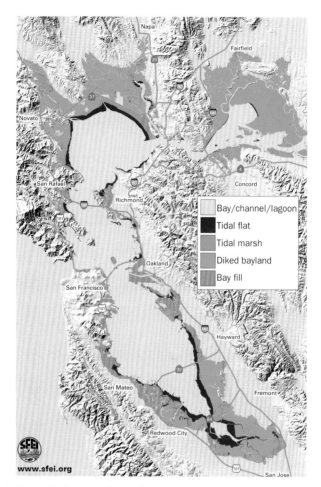

Figure 33b. Current wetlands of the San Francisco Bay.

will encircle the entire bay, giving the public access to the water and restored wetlands. The restoration of San Francisco Bay to health is an inspiring story that involved dozens of organizations and hundreds of individuals over the past four decades. Though there is still work to be done, especially in regard to invisible chemical pollutants, we can look at the bay proudly for what has been accomplished.

The largest remaining tracts of wetlands in the Bay Area are at its eastern edge in Suisun Bay and the Delta, where the Sacramento and San Joaquin Rivers join. Here you can go back 100 years or so to a time when San Francisco Bay was surrounded by such wetlands, a time when the migratory season brought millions of waterfowl through the Bay Area. Montezuma Slough winds through a vast area of marshes and ponds. At very high tides, the marshes are flooded; at low tides the stands of pickleweed *(Salicornia pacifica)* stretch from slough to slough. The area is a delight to fishermen, birders, kayakers and canoeists, duck hunters, and others who seek a respite from the urban Bay Area. In the marshes of the Grizzly Island National Wildlife Refuge (map 26) you can get a sense of what the Delta might have looked liked before it was harnessed to agriculture.

The South Bay is also rich in wetlands, from the Palo Alto Baylands (pl. 57) on the west to the San Francisco Bay National Wildlife Refuge on the east. Although much of this area was once

Plate 59. Pile of salt at the Cargill salt plant in Fremont, 1986.

diked for salt production, thousands of acres have been restored and are once again a part of a rich estuarine ecosystem. The shallow waters of the southern end of the bay combined with high summer temperatures favored salt production there by Native Americans in prehistoric times, and by modern industry. The multicolored ponds that travelers see on approach to the Bay Area's airports are diked salt ponds, colored by organisms that live in the highly saline water. It takes 7 years to evaporate the water to produce salt. The bright white mound near the water's edge in Fremont (pl. 59) is harvested salt waiting to be processed for your dinner table and for industry.

The Islands in the Bay: The Drowned Landscape

Adding visual and geologic interest to San Francisco Bay are one large and several smaller islands: Angel Island, Yerba Buena Island, Alcatraz, Brooks Island, Red Rock, the Brothers and Sisters, and the Marin Islands. They are the tops of hills or high ridges that were flooded as the waters rose in the bay after the last ice age. Belvedere was almost an island, connected to Tiburon Peninsula only by a sand spit flooded at high tide before a causeway was built. Only Treasure Island is an artificial island.

All of the islands consist of Franciscan rocks of several different terranes (see chapter 3). Angel Island (pl. 60), the largest (at 1 square mile) and highest of the bay islands, is also the most geologically interesting. It has a wide variety of Franciscan rock types of the Yolla Bolly Terrane, including unusual high-grade metamorphic rocks. More than two-thirds of the rock exposed on Angel Island is altered graywacke, which was quarried in the 1860s for building material for San Francisco. Franciscan chert and pillow basalt and serpentinite of the Coast Range Ophiolite can also be seen along the island's trails. Before the Sacramento and San Joaquin Rivers cut Raccoon Strait, Angel Island was a continuation of the Tiburon Peninsula ridge (map 14), which has similar rocks.

The other islands consist mostly of the more common Franciscan rocks. Bedrock on Alcatraz (pls. 61, 62) is entirely unal-

Plate 60. Angel Island from Golden Gate Bridge.

Plate 61. Alcatraz from Golden Gate Bridge; Mount Diablo and North Peak in background.

tered graywacke, the same as that at Telegraph Hill in San Francisco. One of the 11 Franciscan terranes—the Alcatraz Terrane—was defined on the basis of the graywacke found on the island. Yerba Buena Island (pl. 63) is largely graywacke, also of the Alcatraz Terrane, with some chert on the west side near the tun-

Plate 62. Graywacke of Alcatraz Terrane on west side of Alcatraz.

Plate 63. Yerba Buena Island, looking east from San Francisco.

Plate 64.
Tilted layers of
graywacke and
shale on west
side of Yerba
Buena Island.

nel entrance. From a boat you can see that the graywacke and
shale of Yerba Buena Island are layered and complexly folded (pl.
64). Treasure Island, the artificial northern extension of Yerba
Buena, was constructed in the 1930s by filling a naturally shallow
part of the bay to provide a site for the 1939 World's Fair commemorating the opening of the Golden Gate and Bay Bridges.

Red Rock (pl. 65), an interesting little island just south of the
Richmond–San Rafael Bridge, has the distinction of being the
meeting place of three counties—San Francisco, Marin, and
Contra Costa. From its name you can probably guess that the
chief type of rock is Franciscan radiolarian chert of the Marin
Headlands Terrane. A deep cave on the island is the site of an attempt to mine manganese. The Marin Islands in San Rafael Bay
and the Brothers and Sisters near the Contra Costa shore are also

Plate 65. Red Rock.

the tops of ridges drowned by rising sea level. The rocks of these islands are predominantly Franciscan graywacke of the Novato Quarry Terrane, with minor amounts of shale, conglomerate, and chert on the Marin Islands.

THE PENINSULA AFFORDS Bay Area residents many splendid opportunities to explore the spectacular beaches, scenery, and rocks of San Mateo County (map 15). Three very different geologic landscapes call to the visitor: the narrow coastline with magnificent tidepools and sandy beaches, the mountains that form the forested backbone of the Peninsula, and the gently sloping urbanized plain and wetlands along San Francisco Bay. The San Andreas Fault cuts a broad swath through the Peninsula and occasionally sends earthquakes rattling across the Bay Area. Only a few roads cross from coast to bay, giving much of the Peninsula a sense of remoteness that belies its proximity to major Bay Area urban centers.

Much of the Peninsula landscape is mountainous. Montara Mountain rises abruptly from sea level to almost 2,000 feet at Pacifica; to the south lie the Santa Cruz Mountains. Skyline Blvd. winds down their crest, rising gradually to 2,500 feet and staying at about that elevation to Santa Cruz County. The Santa Cruz Mountains are highest where the San Andreas Fault makes a bend southwest of San Jose. The highest peaks are Mount Bielawski (3,231 feet) in Santa Cruz County and Mount Umunhum (3,484 feet) and Loma Prieta (3,806 feet) in Santa Clara County (see chapter 8). These mountains are young and still growing.

Geologically, the Peninsula has a split personality (map 16). Bisected by the San Andreas Fault, the western part is attached to the Pacific Plate and is moving slowly but inexorably toward Alaska (see chapter 2). East of the San Andreas Fault, the Peninsula is attached to the North American Plate. The Peninsula landscape owes much of its scenic beauty to active tectonics along several faults that are part of the San Andreas Fault System. All the rocks of the Peninsula have been rearranged extensively by faulting.

The Coastal Landscape

Before looking at the Peninsula's complex geologic mosaic, we should explore the coastal landscape and the processes that have shaped it. The spectacular scenery of the Peninsula's long shoreline is created by the same basic geologic processes that operate elsewhere. Tectonic movements uplift the land or move it laterally along faults; weathering and erosion wear the high places down. Two additional factors are important on the coast: wave action and changes in sea level. Waves crashing ceaselessly against the shore are major shapers of this landscape—battering sea cliffs; forming sea stacks, caves, and arches; moving sand on and off the beaches. Tectonic forces, wave action, and changes in sea level have combined to produce a dramatic and scenic landscape.

The Land Uplifted

Traveling down the Peninsula along the coast, you negotiate the tight curves of Devils Slide on Montara Mountain with its steep drop to the ocean and gratefully take in the gentler view south to the wide, flat bench of Half Moon Bay. Much of the Peninsula shoreline consists of these flats, which are marine terraces that have been raised far above the reach of today's waves by movement on nearby faults (pl. 66). The terraces were carved thousands of years ago as waves crashed against the shore forming wave-cut platforms. Then the platforms were uplifted by tectonic activity (see chapter 1). The Half Moon Bay terrace was cut during one or both high stands of the sea at 105,000 and 82,000 years ago. It is testimony to rapid geologic changes along this coast.

Along much of the Peninsula, uplift of terraces has occurred

Plate 66. Marine terrace south of Half Moon Bay.

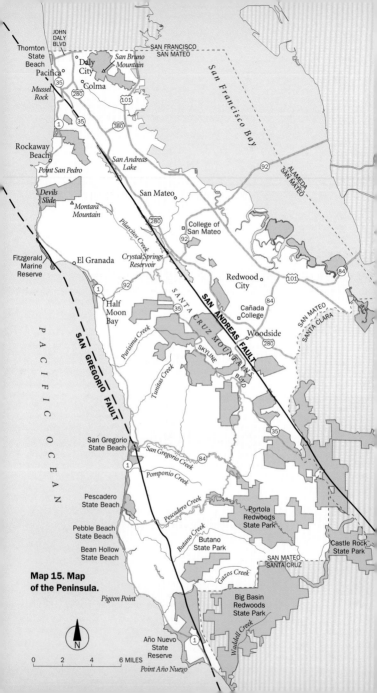

Map 15. Map of the Peninsula.

0   2   4   6 MILES

Map 16.
Geologic map
of the Peninsula.

Artificial fill
Quaternary alluvium
Quaternary bay mud, marsh
Quaternary beach and dune sand
Quaternary marine, terrestrial sedimentary rocks
Quaternary marine terrace deposits
Quaternary/Tertiary sedimentary rocks
Upper Tertiary sedimentary rocks
Lower Tertiary sedimentary rocks

Tertiary volcanic rocks
Franciscan Complex coherent rocks
Franciscan Complex mélange
Great Valley Sequence
Serpentinite, other Coast Range Ophiolite
Salinian Complex granitic rocks

SAN FRANCISCO
SAN MATEO

ALAMEDA
SAN MATEO

San Francisco Bay

SAN MATEO
SANTA CLARA

SAN MATEO
SANTA CRUZ

PACIFIC OCEAN

SEAL COVE FAULT

PILARCITOS FAULT

SAN ANDREAS FAULT

SAN GREGORIO FAULT

2   4   6 MILES

Plate 67. Terraces near Davenport in Santa Cruz County. Several higher terraces to the right are not shown.

many times. Progressively older and more eroded levels of terraces march up the hillsides. On Hwy. 1 you dip down repeatedly from one level to a lower one as you wind down the coast, then climb up again to the next-higher terrace. Flat-topped ridges to the east mark remnants of older terraces. A spectacular flight of six terraces, formed in the past half a million years, can be seen between Santa Cruz and Davenport in Santa Cruz County (pl. 67).

Not all marine terraces are flat lying. The Half Moon Bay terrace has been folded into a gentle syncline, and Hwy. 1 lies on a southward-tilted marine terrace from Mussel Rock in Daly City south to Point San Pedro. The northern end, at the Mussel Rock Transfer Station, is 160 feet above sea level; the southern end dips below sea level at Laguna Salada, 3 miles to the south.

Along the shore, waves today are forming new wave-cut platforms. The wave-cut platform at Fitzgerald Marine Reserve has hosted generations of schoolchildren observing treasures in its tide pools. It is cut into the soft marine sandstone and shale of the Purisima Formation. At a very low tide, you can see a spectacular syncline in the platform where movement along the Seal Cove Fault has folded the layers of rock (pl. 68).

Plate 68. Syncline on wave-cut platform at Fitzgerald Marine Reserve.

Down to the Sea

Although most geologic processes are slow by human standards, along the shore one can see major change over a single lifetime. The interaction of human activity and geologic processes is dramatically clear along the Peninsula coast. Much money has been invested in efforts, ultimately futile, to keep a railroad, a highway, backyards, and houses from crashing to the beach below. Many areas along this coast consist of young and soft sediments that are particularly prone to erosion, especially where they are undercut by waves. Older and highly fractured rock at Devils Slide on Hwy. 1 is notorious for road washouts, which can affect the coastal communities of the central Peninsula for months at a time.

Coastal erosion has been a problem here for more than a century. Imagine trying to maintain a railroad across soft and fractured rock under attack by the waves. In 1905 construction started on the Ocean Shore Railroad to link San Francisco with Santa Cruz and to carry weekend bathers to the beaches at Pacifica, El Granada, and Half Moon Bay. Maintaining tracks in loose sediment was very difficult, and the project was finally abandoned in 1921 when the company went bankrupt. A few portions of the railroad bed are still visible along the shore. In 1939 the California Department of Highways (now Caltrans) took over the right-of-way and used it for Hwy. 1. This stretch of Hwy. 1

eventually suffered the same fate as the railroad. In stormy years, landslides in the soft sediment sent pieces of pavement sliding down the cliffs. The 1957 Daly City earthquake damaged much of the highway, and efforts to maintain it were dropped.

Thornton State Beach at the west end of John Daly Blvd. in Daly City was once a lovely little park with picnic tables sheltered from the winds and easy access to the beach (pl. 69, *top*). Heavy rain in the El Niño storms of the 1982/3 and 1983/4 winters washed out the road and sent young sediments of the Merced and

Plate 69. Thornton State Beach. *Top*, 1976. *Bottom*, 1999. Note that the access road is now gone.

Colma Formations flowing down the cliff. The slides buried the picnic area in sand and destroyed both the park and road (pl. 69, *bottom*).

Homes built on the bluffs to take advantage of magnificent ocean views have been sliding since the 1950s, when the hills were leveled for the Westlake Palisades subdivision (pl. 70, *top*). Above

Plate 70. *Top,* Daly City in 1953 before development. *Bottom,* After construction of Westlake Palisades in 1973. Notice the empty lots at the edge of the landslide right of center.

Plate 71. Avalon Canyon landslide, December 2003. This photo was taken two weeks after the landslide.

Mussel Rock, homes constructed at the edge of a huge landslide where the San Andreas Fault goes to sea soon began to slide downhill. Eleven homes were moved inland within months of their construction (pl. 70, *bottom*). The inexorable erosion continued, and each El Niño season threatened more homes in this area. In early 2000, 21 homes were moved after geologic investigations showed that the landslide and cliff retreat could not be stabilized. Local residents have paid dearly for their dramatic views of the ocean. The cliffs retreated again in 2003 when a large landslide in nearby Avalon Canyon sent about 500,000 cubic yards of earth flowing out into the ocean (pl. 71). By the end of the winter, the waves had removed all the landslide debris.

Faults That Carve Up the Peninsula

The San Andreas and its associated faults are major players in Peninsula geology. Three large faults of the San Andreas Fault System cut the Peninsula: the San Andreas, the Pilarcitos, and the San Gregorio Faults (map 17). The dominant movement on all three faults is strike-slip (see chapter 2), but many smaller thrust faults are also present, especially at the southern end of the Peninsula where the San Andreas Fault makes a slight bend. We know that movement on some of these faults is very young, because rocks that are less than 1 million years old have been folded and faulted, and marine terraces younger than 100,000 years are uplifted and tilted along the coast.

These faults have left their impact on the landscape and rocks. The broad valley of the San Andreas Fault bisects the Peninsula from southeast to northwest (pl. 11). The valley is almost a mile wide at Crystal Springs Reservoir. San Andreas Lake, at the valley's northern end, is one of several lakes or sag ponds in the fault zone. After the 1906 earthquake, the significance of the valley was recognized and the newly "discovered" fault was named after the lake. The dams that impound San Andreas and Crystal Springs Reservoirs were not destroyed in the 1906 earthquake, although part of the Upper Crystal Springs Dam shifted as much as 8 feet.

The Pilarcitos Fault is an older branch of the San Andreas system. It extends from the coast near Rockaway Beach about 26 miles southeast to its merger with the San Andreas Fault near the Santa Cruz county line. The Pilarcitos, which is no longer active, was probably the dominant fault on the Peninsula before the San Andreas Fault became active here 3 or 4 million years ago. The San Gregorio Fault, another major fault of the San Andreas system, lies along the western edge of the Peninsula. It lies partly under the ocean but crosses land at two locations, at Half Moon Bay, where it is called the Seal Cove Fault, and to the south from San Gregorio State Beach to Point Año Nuevo. It is still active and ruptured as recently as about 650 years ago, when it cut through and offset part of an archaeological site west of the Half Moon Bay airport. Although its movement is dominantly strike-slip, locally it has uplifted segments of the coast. Looking west from Hwy. 1 across the Half Moon Bay airport, you can see the Seal Cove Bluff, an east-facing fault scarp (cliff) formed by uplift of the Pillar Point area (pl. 72).

The San Andreas, Pilarcitos, and San Gregorio Faults divide the Peninsula into four geologic units (table 3). Three of them, the Pigeon Point, La Honda, and Pilarcitos Blocks, are on the Pacific Plate. One, the San Francisco Bay Block, lies east of the San Andreas Fault. Each block has different rocks, a different sequence of the same rocks, or a different geologic history than the others. Each block has been carried northward to the Peninsula from its place of origin to the south by strike-slip fault move-

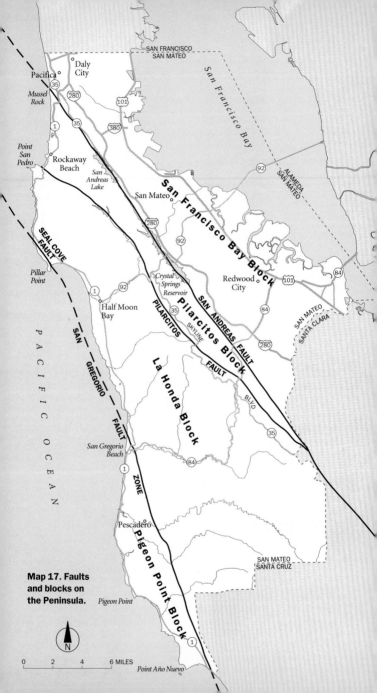

SAN FRANCISCO
SAN MATEO

Daly
City

Pacifica

Mussel
Rock

Point
San
Pedro

Rockaway
Beach

San
Andreas
Lake

San Mateo

SEAL COVE
FAULT

Pillar
Point

Half Moon
Bay

Crystal
Springs
Reservoir

Redwood
City

SAN ANDREAS FAULT

PILARCITOS

Pilarcitos Block

SKYLINE

FAULT

La
Honda
Block

BLVD.

San Gregorio
Beach

ZONE

Pescadero

Pigeon
Point Block

Pigeon Point

San Francisco Bay

SAN FRANCISCO BAY BLOCK

SAN GREGORIO FAULT

PACIFIC OCEAN

San Francisco Bay Block

ALAMEDA
SAN MATEO

SAN MATEO
SANTA CLARA

SAN MATEO
SANTA CRUZ

Map 17. Faults
and blocks on
the Peninsula.

N

0   2   4   6 MILES

Point Año Nuevo

Plate 72. Seal Cove Bluff, beyond the airport, uplifted by movement on the Seal Cove Fault.

TABLE 3. Blocks and Rock Types on the Peninsula

Blocks	Basement Rocks
On the Pacific Plate	
Pigeon Point	Great Valley Complex–Pigeon Point Formation
La Honda	Salinian Complex–Granitic rocks
Pilarcitos	Franciscan Complex–Permanente Terrane; Great Valley Complex
On the North American Plate	
San Francisco Bay	Franciscan Complex–San Bruno Mountain and Permanente Terranes, mélange

ment. Thus, the Peninsula has been assembled by faulting over a long period of time that has thoroughly rearranged the older rocks. Table 3 lists the blocks and rock types mentioned in the text. Because the geologic picture is so complex, we discuss only the major types of rocks.

The Basement Rocks

All three types of basement (oldest) rocks found in the Bay Area are present on the Peninsula (map 16): the granitic rocks of the Salinian Complex on the La Honda Block, the Franciscan

Complex on the Pilarcitos and San Francisco Bay Blocks, and the Great Valley Complex on the Pigeon Point and Pilarcitos Blocks (see chapter 3 for descriptions of these rocks). This sounds simple, but, in fact, the picture on the Peninsula is quite complicated. For example, although north of the Golden Gate, Franciscan basement rocks occur only east of the San Andreas Fault, that is not the case on the Peninsula. A wedge of Franciscan rocks, on the Pilarcitos Block, is present between the San Andreas and Pilarcitos Faults from Pacifica to the Santa Clara county line (map 16). This wedge was captured by the Pacific Plate several million years ago and is now traveling with it. This is the only place in the Bay Area where Franciscan rocks are found west of the San Andreas Fault.

Pigeon Point Block

West of the San Gregorio Fault, the basement rocks are interesting sandstones and conglomerates of the Upper Cretaceous Pigeon Point Formation, a marine sedimentary rock that underlies the marine terrace deposits between Pescadero State Beach and Point Año Nuevo (map 17). These rocks were formerly thought to be part of the Salinian Complex, but recent work has shown that they correlate with the Atascadero Formation of the Great Valley Sequence, part of the Great Valley Complex. The Atascadero Formation is found about 100 miles to the south in southern Monterey County.

The Pigeon Point Formation was deposited by turbidity flows, and many features typical of such rocks can be seen at low tide in spectacular outcrops, especially at Pebble Beach, south of Pescadero State Beach, and Bean Hollow Beach (pl. 73). Each separate turbidite is graded; that is, it has coarse sandstone or pebble conglomerate at the base, grading into increasingly finer sediments above (pl. 20). The individual turbidite layers are often separated by a layer of shale formed from fine particles that settled on the seafloor between turbidity flows. Some of the turbidites have been eroded into a fretwork known as tafoni, or honeycomb weathering. Two mechanisms, not mutually exclusive, have been proposed for its formation: chemical interaction of salt spray with the sandstone and differential erosion of more weakly cemented portions of the rock.

On several Peninsula beaches the sandstone cobbles contain round holes, like Swiss cheese (pl. 74). The rock was not formed

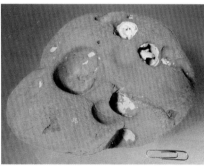

Plate 73 *(above)*. Pigeon Point Formation turbidites at Pebble Beach. Note the alternating layers of sandstone and shale and the fretwork (tafoni; lower right).

Plate 74 *(left)*. Pholad-bored rock with pholad clam.

this way; the holes are made by rock-boring clams called pholads. One end of their shell is rough, which enables the clam to grind into the rock. If you look carefully, you can see that some of the rocks still have remnants of the pholad shell at the wide end of the hole. The clam starts boring when it is small and makes the hole bigger as it grows, trapping itself inside.

La Honda Block

Between the San Gregorio and Pilarcitos Faults, the basement rocks are the granitic rocks of Montara Mountain, which are about 92 million years old (Cretaceous). They can be seen in road cuts on Hwy. 1 just south of Devils Slide (pl. 2). A few patches of even older rocks, such as marble and other metamorphic rocks, are interspersed with the granitics. The marble has been mined for cement at Montara. The Montara granitics are highly fractured by their long travels with the nearby San Andreas Fault System. They formed far to the south during the Mesozoic subduction and have traveled northward with the Pacific Plate for millions of years (see chapter 2). They are similar to granitic rocks on Bodega Head and Point Reyes that have been offset northward by movement on the San Andreas Fault System.

Pilarcitos Block

Between the Pilarcitos and San Andreas Faults, the basement rocks are not granitic; they are predominantly Franciscan. When fault movement jumped eastward from the Pilarcitos to the San Andreas a few million years ago, a piece of the Franciscan Permanente Terrane was left behind on the Pacific Plate. At this location the Permanente Terrane (table 2) includes graywacke, pillow basalt, minor amounts of chert, and a limestone called the Calera Limestone (pl. 75). Microfossils in the limestone indicate that these rocks were formed near the equator (at about 22°N) in a deep ocean environment, probably on a seamount or submarine plateau. You can examine the Calera Limestone at close quarters

Plate 75. Calero Limestone in Pacifica Quarry at Rockaway Beach.

in the former Pacifica Quarry at Rockaway Beach, which is now part of the Golden Gate National Recreation Area.

To add to the complexity of Peninsula geology, a very small bit of Great Valley Complex has been faulted into the Pilarcitos Block and is present near the intersection of Hwy. 92 and Skyline Blvd. Although the outcrop is too small to show on the geologic map, this unit is important because it includes a conglomerate containing distinctive molluskan fossils and is like rocks found 135 miles north along the coast in southern Mendocino County, which have been carried northward by fault movement.

San Francisco Bay Block

The basement rocks east of the San Andreas Fault are Franciscan. San Bruno Mountain (pl. 76), the geographic feature that gives its name to the San Bruno Mountain Terrane (table 2), consists entirely of a very large block of graywacke and shale. The graywacke was formed in the same way as other Franciscan graywackes—as turbidite deposits in a deep marine trench associated with subduction (see chapter 2). This graywacke was formed by many individual flows, some massive, others of varying thickness (pl. 54).

South of San Bruno Mountain, the Franciscan basement consists of the Permanente Terrane and mélange (see chapter 3).

Plate 76. Aerial photo of San Bruno Mountain.

Plate 77. Serpentinite exposed along Hwy. I-280.

Blocks of greenstone, chert, metamorphic rocks, and serpentinite are scattered throughout the mélange in this area. The College of San Mateo is built on mélange; greenstone and serpentinite underlie Cañada College. The characteristics of a serpentinite landscape are particularly well exposed along I-280 from San Andreas Lake south to Hwy. 92 (pl. 77). Here you can see a typical crushed serpentinite mélange matrix with random blocks of more resistant serpentinite scattered through it. The soil cover is thin, and in spring California poppies provide a beautiful contrast to the blue green serpentinite.

A Record of Past Seas

Overlying the basement rocks on both sides of the San Andreas Fault is a thick sequence of Tertiary marine sedimentary rocks, including sandstone, shale, mudstone, and conglomerate, and some volcanic rocks, ranging in age from Paleocene to Pliocene (65 to 1.8 million years). They provide a record of the marine basins that formed after subduction ended (see chapter 2) and the change from deep ocean to shallow marine conditions. Along the coast at Montara Mountain, Paleocene sedimentary rocks are exposed in road cuts (pl. 78), where they form the weak sandstone and shale layers that are causing so much trouble at Devils Slide. Although originally deposited as flat-lying layers of sand

Plate 78. Tilted sandstone and shale layers at Devils Slide.

and mud on the ocean floor, they have been tilted up almost vertically by tectonic forces. The same rock layers make up nearby Point San Pedro, which one day will be a sea stack (pl. 79). South of Montara Mountain, the La Honda Block consists of a thick sequence of Tertiary marine sedimentary and volcanic rocks, described below. On most of the Peninsula, the sedimentary rocks are covered with magnificent oak-madrone or redwood forest and are poorly exposed. Skyline Blvd. south of Hwy. 92 crosses several of the units, but outcrops are sparse. The rocks tend to be somewhat similar in appearance, and it is not easy to tell one unit from another. A good place to see an example of these rocks is at Castle Rock State Park, off Skyline Blvd. just across the Santa Cruz county line. Here, the Vaqueros Sandstone makes prominent outcrops that are favorites of rock climbers (pl. 80). Much of the sandstone has been weathered by chemical processes into

Plate 79. Point San Pedro showing sandstone and shale layers. *Top,*
View with old roadbed of the Ocean Shore Railroad running along cliff
of Montara Mountain (center right). *Bottom,* Close-up of Point San Pedro.

tafoni, cavelike holes and a delicate honeycomb pattern on the
rock.

The harder sedimentary rocks, such as the lower Tertiary Vaqueros and Butano Sandstones and the Whiskey Hill Formation,
are more resistant to erosion and form the ridges along the spine
of the Peninsula. One of the most resistant of these rocks, the
sandstone of the Whiskey Hill Formation, is an especially prominent unit on the Peninsula. It underlies Kings Mountain Ridge to
the southwest of Crystal Springs Reservoir and part of Jasper

Plate 80. Vaqueros Sandstone, Castle Rock State Park, Santa Cruz County.

Ridge, south of Woodside. The light-colored Butano Sandstone underlies the hills west of the Hwy. 84–Skyline Blvd. intersection and the ridge southwest of Portola State Park. The Butano Sandstone provides important evidence for movement on the San Andreas Fault System. The sandstone grains in it are very similar to those in the Point of Rocks Sandstone found in the Coast Ranges almost 200 miles to the south. The units were deposited in the same marine basin 40 to 50 million years ago and subsequently were split apart by movement on the San Andreas Fault System. The Butano Sandstone was attached to and carried northward with the Salinian Block. This match across the fault is one line of evidence for the total amount of offset across the San Andreas Fault (fig. 13). Similarly, the Santa Cruz Mudstone, a younger sedimentary rock of late Miocene age and a common unit on the Peninsula, was deposited in a marine basin split by the San Gregorio Fault. Part of the basin sediments are now found about 70 to 90 miles to the northwest at Point Reyes.

A common rock on the Peninsula is the sedimentary Purisima Formation of Pliocene age. The Purisima is subdivided into several units, some relatively resistant to erosion, but many weak and easily washed away, as can be seen in road cuts along Hwy. 1 from Moss Beach to Pescadero State Beach (pl. 81). These

Plate 81. Gullying in Purisima Formation beneath terrace deposits in road cut along Hwy. 1.

rocks closely match rocks at Point Reyes formerly called the Drakes Bay Formation; they too have been offset northward by the San Gregorio Fault.

Fossils are common in the Purisima Formation, and one of the best places to see them is at San Gregorio State Beach. In winter, when there is less sand on the beach and more rocks are exposed, you can see fossil clams and burrows in the rocks along the trail to the beach (pl. 82). The burrows, probably formed by worms, clams, or shrimp, are particularly well exposed in the ceiling of a cave (which can only be reached at low tide) along the beach just to the north of the trail. These fossils indicate that the Purisima was deposited under shallow marine conditions. At very low tides you can walk far enough west to see a thick bed of tuff (volcanic ash) exposed in the cliffs north of the trail (pl. 83). It is about 2.6 million years old, and its source was probably an eruption near Lassen Peak in northeastern California.

Outcrops of two Tertiary volcanic rocks occur on the Peninsula. The Mindego Basalt, which is about 20 million years old, lies west of the San Andreas Fault. Associated sedimentary rocks that contain fossil mollusks and microorganisms indicate that it probably erupted from submarine vents into a marine basin formed along the early San Andreas Fault System. The Mindego

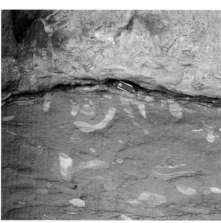

Plate 82. *Top,* Fossil clams in Purisima Formation along trail to beach, San Gregorio State Beach. *Bottom,* Burrows in Purisima Formation at San Gregorio State Beach.

Basalt probably erupted south of the Bay Area and was carried northward to the Bay Area by fault movement. A part of the Mindego Basalt traveled about 125 miles farther north, to Mendocino County by movement along the San Gregorio Fault. There it is called the Iversen Basalt.

The Page Mill Basalt, which is about 15 million years old, lies east of the San Andreas Fault. Its eruption probably marks the beginning of the Peninsula branch of the San Andreas Fault.

Plate 83. Ash layer (thin white upper layer) at San Gregorio State Beach.

The Youngest Rocks of the Peninsula

Beneath the urbanized eastern foothills of the Santa Cruz Mountains and bay plain are several young (Quaternary) marine and terrestrial sedimentary rocks, the Colma, Merced, and Santa Clara Formations (map 16). The main outcrops of the Santa Clara Formation are in Santa Clara County, and they are described in chapter 8. These young rocks are geologically important because they record both changes in sea level and tectonic activity along the San Andreas Fault System over the past several million years. The Colma and Merced rocks are poorly consolidated; that is, the grains are not well cemented, and the rock is soft and easily eroded. Most of the landslides along the coast south of San Francisco have occurred in them.

The Merced Formation is a treasure trove of interesting geology; it records evidence of sea level changes, marine life, fault movement, and distant volcanic eruptions. Look closely at the Merced and you can see a variety of fossils, both marine and terrestrial. Spectacular exposures in the sea cliffs between Fort Funston in San Francisco and Mussel Rock in Daly City (pl. 84) provide a glimpse into the dynamic interaction of San Andreas Fault System tectonics and changes in sea level through time. The

Plate 84. Layers of sand and pebbles in the Merced Formation at Thornton Beach, Daly City.

Merced Formation, which was deposited from 2 or 3 million years ago to about half a million years ago, records the long-term change from deep to shallow water as this coastal area was uplifted by faulting. The older Merced sediments at Mussel Rock were deposited in deeper continental shelf environments than the younger, shallow-water Merced rocks to the north near Fort Funston. A walk south from Ocean Beach along the cliffs starts in the younger deposits and goes back in time to the older and deeper deposits.

The Merced rocks also record shorter sequences of changes in relative sea level. At the base of each short sequence are marine sands deposited in an offshore environment, representing a time when sea level was high. As sea level fell, shallow marine sands were deposited over the deeper-water sediments. As sea level continued to fall, this area emerged from the sea, beach and dune sands formed, and finally pond, swamp, and marsh sediments were laid down. These repeating cycles can be seen in the rocks all along the cliffs. The cycles may be related solely to changes of sea level due to worldwide climate fluctuations or also to tectonic movements along the San Andreas Fault Zone.

Two prominent layers of volcanic ash occur in the Merced sediments. Volcanic glass in the lower part of the formation

Plate 85. Rockland Ash, the prominent white layer, in the Merced Formation in cliff below Fort Funston. Note tilt of Merced layers.

comes from a huge eruption that occurred about 774,000 years ago at Long Valley in eastern California. The other ash occurs in the upper part of the Merced (pl. 85). It has been identified as the Rockland Ash, which erupted from a volcano near Lassen Peak about 570,000 years ago. These layers of ash provide a framework for a very important event in Bay Area geology that is recorded in the rocks of the Merced Formation. About 960 feet below the top of the Merced and 380 feet below the Rockland Ash, the mineral composition of the sandstone changes. In the lower two-thirds of the Merced below the ash, mineral grains in the sandstone reflect erosion from local sources, primarily Franciscan rocks. Above the ash, mineral grains from Sierra Nevada rocks are present in the sediments. This change, which occurs in sediments that are about 620,000 years old, marks the beginning of drainage from the Central Valley through the Bay Area (see chapter 6).

Plate 86. Colma Formation at Thornton Beach. Note dark soil layer at top.

Overlying the Merced Formation are the younger sedimentary deposits of the Colma Formation (pl. 86). On the Peninsula the Colma Formation, like the Merced, consists of sand deposited in coastal marine and estuarine environments. Before the Colma was deposited on top of the Merced Formation, the Merced rocks were tilted and folded. The nearly flat-lying Colma layers rest on tilted Merced layers. Most of the tilting must have stopped before Colma deposition began, because otherwise the Colma would also be tilted. The Colma Formation was deposited during the last interglacial period, about 80,000 to 125,000 years ago, along the coast and in a narrow seaway that extended diagonally across the Peninsula between the ocean and the bay, where the city of Colma is today (fig. 32). Both the Colma and Merced Formations have been uplifted very recently to several hundred feet above sea level, a reminder that the Bay Area landscape is being shaped actively today by tectonic forces.

FOR MOST BAY AREA RESIDENTS, mention of the South Bay calls up a picture of the urbanized bay plain at the southern end of San Francisco Bay: Silicon Valley, a web of interlaced freeways, heat, and smog. But get past the urban sprawl and you find exceptionally interesting geology never suspected by freeway drivers. Beyond the cityscape are rolling hills and mountains, rivers and reservoirs, ranches, horses, grapes, and extensive open space (map 18). Just a few minutes off any freeway can take you on a trip into the past, geologically and culturally. The many parks in the area provide excellent access to the varied geology of the South Bay. Here are ancient rocks that tell of long-ago plate collisions, faults creating a dramatic mountainous landscape, and mercury mines once important to the economic health of the entire nation (map 19).

The "South Bay," as referred to in this book, means Santa Clara County, although geographically southern Alameda County and the southern Peninsula are often considered part of the South Bay. The Santa Cruz Mountains on the west and the Diablo Range on the east frame the area, like the letter V pointed south, with the Santa Clara Valley nestled between them. At its southern end, the valley narrows to a corridor between mountains with elevations of 3,000 to 4,000 feet. The highest peaks of the Santa Cruz Mountains are in Santa Clara County, Loma Prieta (3,806 feet) and Mount Umunhum (3,484 feet). Mount Hamilton in the Diablo Range rises to 4,209 feet, but nearby Copernicus Peak, at 4,373 feet, is a little higher. On a clear day Lick Observatory at the top of Mount Hamilton provides a magnificent view of the South Bay (see chapter opening photo). It is an expansive landscape of high mountains, rolling foothills, and wide and narrow valleys, yet all at a scale that can be encompassed in a single vista from the mountaintop.

The Diablo Range marks the eastern border of the Bay Area. It receives much less moisture than the Santa Cruz Mountains, a difference that is clearly shown by the vegetation. A transect from the fog-loving redwoods on the west takes you through oak grasslands, chaparral and grassland, then up into the madrone and oak woodlands of the drier Diablo Range. Much of this hilly South Bay landscape, like most of the Bay Area landscape, is very young, the product of recent tectonic activity along the San Andreas Fault System. Throughout the area, young rocks only 1 or 2 million years old are folded, faulted, and uplifted. Steep canyons and many large landslides in both mountain ranges attest to a recent and rapid uplift.

The Valley That Was:
The Santa Clara Valley

Wheel the family time machine out of the garage and head back to the Santa Clara Valley as it was in 1940 or 1950. Smell the apricot, plum, and cherry blossoms in springtime; fill your fruit basket in summer. Picnic by the clear creeks that flow from the mountains to San Francisco Bay. You are in the "Valley of Heart's Delight," once the fruit basket of the nation (fig. 34). Here, the deep, rich alluvial soil is perfect for fruit trees.

Figure 34. Historic photo of the Santa Clara Valley before development: foothill orchards near Los Gatos.

The Santa Clara Valley is at the southern end of a down-dropped fault block between the San Andreas and Calaveras Faults; San Francisco Bay is the drowned northern end. The valley is filled with up to 1,500 feet with sediments eroded out of the Santa Cruz Mountains and the Diablo Range. Both the uplift of the ranges and the deposition of the sediment have occurred in the last few million years. Erosion of the uplands must have taken place very rapidly to form such thick, young sedimentary deposits in the valley.

text continues on page 192

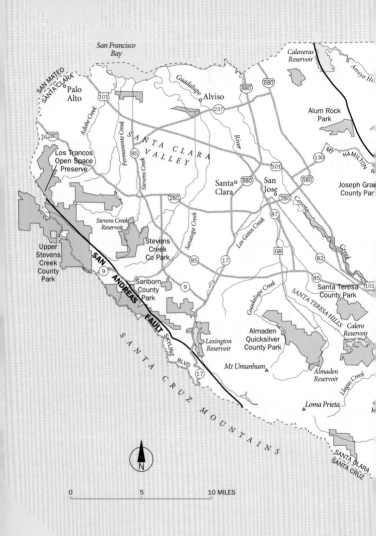

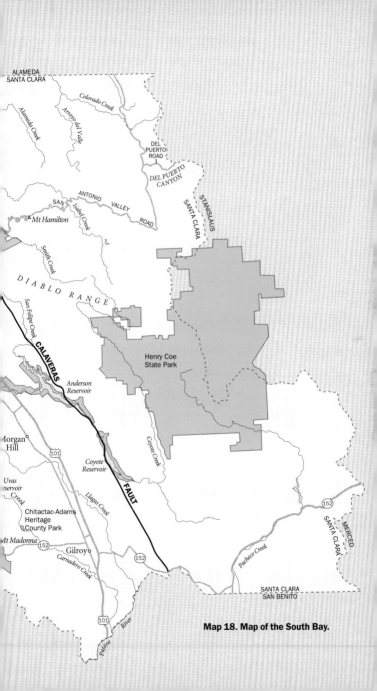

Map 18. Map of the South Bay.

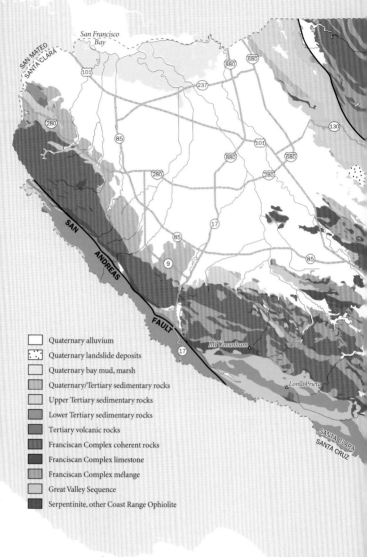

Quaternary alluvium

Quaternary landslide deposits

Quaternary bay mud, marsh

Quaternary/Tertiary sedimentary rocks

Upper Tertiary sedimentary rocks

Lower Tertiary sedimentary rocks

Tertiary volcanic rocks

Franciscan Complex coherent rocks

Franciscan Complex limestone

Franciscan Complex mélange

Great Valley Sequence

Serpentinite, other Coast Range Ophiolite

N

0   5   10 MILES

Map 19. Geologic map of the South Bay.

The unconsolidated sediments beneath the valley consist of many layers of gravel, sand, silt, and clay (fig. 35). Some of these layers are stream deposits; others, particularly some of the clay layers, were formed when earlier San Francisco bays covered the area. Remember that San Francisco Bay has come and gone several times over the past half a million years (see chapter 6). When sea level was higher than today during interglacial times, the bay was larger and at times covered the northern part of Santa Clara Valley. Mud settled to the bottom of the bay. When sea level dropped as glaciers built up to another ice age, the bay waters drained away and streams again deposited their coarser sediment in the valley on top of the muds, which hardened into layers of clay.

The clay layers are fine grained and relatively impervious to water. The coarser sand and gravel layers store and transmit water, forming the important aquifers that have supplied water to orchards and growing populations from the earliest days of valley agriculture. As more and more wells were put in, water levels dropped, and wells were drilled deeper and deeper to tap the

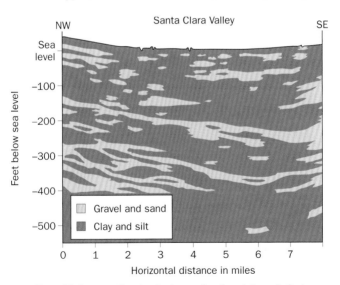

Figure 35. Cross section showing layers of sediments beneath Santa Clara Valley. Gravel and sand aquifers hold water; clay and silt aquitards are impervious.

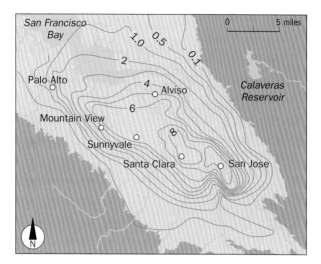

Figure 36. Ground subsidence in the South Bay. Contour lines show areas of equal subsidence (in feet), 1934 through 1967.

lower aquifers. The withdrawal of the ground water led to the compaction of the sediments, which in turn caused the land to subside. The Santa Clara Valley area was the first place in the United States where the problem of subsidence and decline in water table due to the withdrawal of groundwater was recognized, and it was also the first to take remedial action. The earliest efforts to stop the subsidence took place in the 1930s, when the water level had dropped as much as 80 feet. Five reservoirs were built to allow recharge of the water table through slow percolation. The recharge slowed but did not halt the subsidence. Between 1934 and 1967, the land had subsided as much as 8 feet in the San Jose area (fig. 36). By the 1970s, the total subsidence was almost 13 feet, and the water table had declined about 250 feet.

The explosive population growth and industrialization of the 1960s increased the demand for water to such an extent that local water could not meet the need, and water import facilities were built. Today the area depends on water imported from the Sierra Nevada for more than half of its water supply. The rest comes from local groundwater in the valley's aquifers and from surface waters flowing into creeks that fill Santa Clara County's 10 reservoirs before continuing to the bay.

South Bay Faults and Structure

The South Bay, like so much of the Bay Area, is a maze of faults (map 20). The young landscape is shaped both by strike-slip (sliding) motion, related to the movement of the Pacific Plate past the North American Plate, and by thrust faulting that results from the small component of collision between the plates (see chapter 2). On the west the San Andreas Fault slices through the Santa Cruz Mountains near the Santa Cruz–Santa Clara county line. The other major South Bay fault, the Calaveras, branches off the San Andreas near Hollister, south of the Bay Area, and goes northward roughly along the eastern margin of the Santa Clara Valley.

The effects of the strike-slip faults can be seen throughout the landscape in the linear valleys, offset streams, and other common fault-related topographic features (fig. 12). The topographic expression of the San Andreas, which on the Peninsula is a very broad linear valley, in the South Bay is a series of smaller valleys and higher mountains. Upper Stevens Creek and Sanborn County Parks lie in a small rift valley along the San Andreas, and at Sanborn the fault runs through Lake Ranch Reservoir, originally probably a sag pond. On the east the path of the Calaveras Fault is also marked by a series of valleys, including Halls Valley in Joseph Grant County Park. The Calaveras (pl. 87) and Coyote Reservoirs occupy two of the larger valleys.

Plate 87. Calaveras Reservoir in Calaveras Fault Zone, looking north.

Plate 88. View south from near Los Trancos earthquake trail to Loma Prieta mountain (at arrow).

Visit the San Andreas Fault Trail at Los Trancos Open Space Preserve west of Palo Alto (just over the county line in San Mateo County) to get a sense of the fault and its effect on rocks and landscape. Here you can see examples of topographic features, such as the 1906 fault rupture, sag ponds, pressure ridges, and springs. At the first stop you can stand by boulders of conglomerate and look at their source on Loma Prieta, 23 miles southeastward on the other side of the fault (pl. 88). It took 1 or 2 million years of movement to bring those rocks to their present spot.

Major earthquakes have occurred in historic times on both the San Andreas and Calaveras Faults in the South Bay. The most recent was the October 17, 1989, Loma Prieta earthquake of magnitude 6.9 with an epicenter in Santa Cruz County just a few miles west of Loma Prieta Peak. Rocks west of the fault slid past those to the east about 6 feet, and the Santa Cruz Mountains were uplifted about 14 inches. This was the strongest earthquake in the Bay Area since the 1984 Morgan Hill earthquake of magnitude 6.1 and the first near magnitude 7 since the 1906 earthquake.

Thrust and reverse faults (fig. 10) are particularly important in the South Bay and are largely responsible for much of its hilly topography and highest mountains. The effects of transpression,

San Mateo / Santa Clara

101

280

SAN

9

Salinian

ANDREAS

85

FAULT

Block

SIERRA

9

17

AZUL

SIERRA

FAULT

Sierra Azul Block

BERROCAL

SARGENT

FAULT

880

237

680

EVERGREEN

FAULT

WARM SPRINGS FAULT

ARROYO AGUAGUE FAULT

CALAVERAS FAULT

Alum Rock

CLAYTON

SILVER

130

101

CREEK

880

680

280

FAULT

87

280

Silver Cr

17

82

85

SHANNON

FAULT

New Almaden Block

CALERO FAULT

SANTA CLARA / SANTA CRUZ

Santa Clara / Santa Cruz

FAULT

—— Fault with dominantly strike-slip motion

—— Fault with dominantly thrust motion

N

0   5   10 MILES

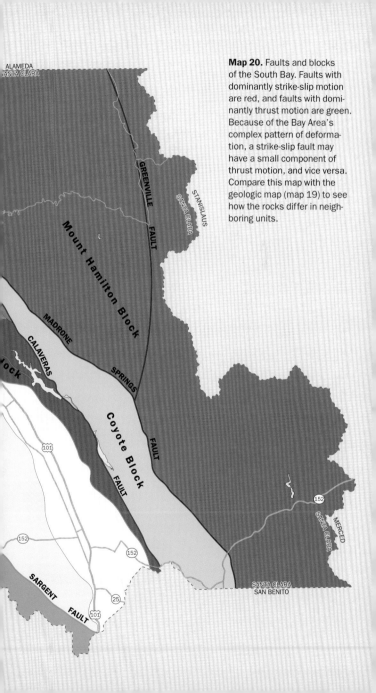

Map 20. Faults and blocks of the South Bay. Faults with dominantly strike-slip motion are red, and faults with dominantly thrust motion are green. Because of the Bay Area's complex pattern of deformation, a strike-slip fault may have a small component of thrust motion, and vice versa. Compare this map with the geologic map (map 19) to see how the rocks differ in neighboring units.

a combination of strike-slip and colliding movement (see chapter 2), are evident as the crust is being shortened by both faulting and folding. Active faults occur on both sides of the Santa Clara Valley—in the foothills of the Santa Cruz Mountains and the Diablo Range. These faults have formed ridges and hills and folded rocks less than a million years old. Thrust faulting is pushing the Santa Cruz Mountains high into the air where the San Andreas Fault makes a slight bend to the west (a leftward bend; fig. 15, *upper left*). The Santa Cruz Mountains are highest in the area of the bend and its related faulting.

The South Bay's Diverse Rocks

A wide diversity of rock types underlies the mountains, rolling hills, and valleys of the South Bay, brought together by the Bay Area's dynamic geologic history. The oldest rocks, the basement rocks, are the Franciscan and Great Valley Complex rocks formed during the Mesozoic plate collision (see chapter 2). Overlying them are thick layers of Tertiary sedimentary and volcanic rocks and widespread outcrops of younger sedimentary deposits that have not turned into rock yet. All except the very youngest have been moved and rearranged extensively by movement on the San Andreas Fault System.

The San Andreas, Calaveras, and other faults divide the South Bay into seven packets or blocks (map 20), each with a more or less different sequence of rocks and geologic history (table 4) and each separated from its neighbors by faults, both active and ancient. A small part of the Salinian Block lies in Santa Clara County west of the San Andreas Fault. Five elongate packets lie between the San Andreas Fault and the Diablo Range on the east: the New Almaden, Sierra Azul, Silver Creek, Alum Rock, and Coyote Blocks. The Mount Hamilton Block, east of the Calaveras Fault in the Diablo Range, is the largest packet; only its western half lies in Santa Clara County.

The rocks *within* each packet are related in origin, but the individual packets have been moved around to such an extent that neighboring rocks may be very different, as you can see by comparing map 20 with map 19. Movement along these many faults has brought the blocks together over time to produce the South

TABLE 4. Terranes and Blocks in the South Bay

Blocks	Basement Rocks	Terranes
On the Pacific Plate		
Salinian Block	Salinian Complex	Salinia
On the North American Plate		
Sierra Azul	Great Valley Complex	Healdsburg, Del Puerto
New Almaden	Franciscan Complex	Permanente, Marin Headlands, mélange
Silver Creek	Franciscan Complex	Mélange
	Great Valley Complex	Healdsburg, Del Puerto
Alum Rock	Franciscan Complex	Mélange
	Great Valley Complex	Healdsburg, Del Puerto
Coyote	Great Valley Complex	Del Puerto
Mount Hamilton	Franciscan Complex	Yolla Bolly, Burnt Hills, mélange

Bay's geologic complexity. Remembering that rocks in each packet have moved relative to neighboring packets can help you understand why, for example, Great Valley Sequence and Franciscan rocks are juxtaposed here.

The Basement Rocks

Of the three types of basement rocks found in the Bay Area (see chapter 3 for a description of these rocks), two—the Franciscan and Great Valley Complexes—are common in Santa Clara County (map 19). The third—the Salinian Complex—underlies the small area of the county that is west of the San Andreas Fault, but its granitic rocks do not occur at the surface in the county. Franciscan rocks underlie most of the Santa Cruz Mountain foothills on the west and form the core of the Diablo Range on the east. The Diablo Range is a great anticline (fig. 14) with Franciscan rocks in its core and extensive outcrops of the Great Valley Complex on either side (map 21). The Great Valley rocks that once covered the Franciscan core have been eroded away. The western part of the anticline lies in Santa Clara County; the east-

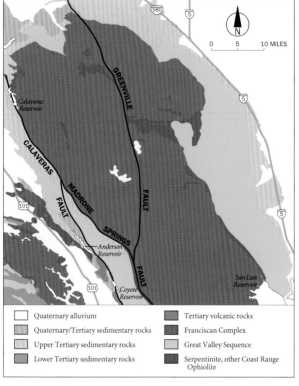

Map 21. Geologic map of the Diablo Range.

Legend:

- Quaternary alluvium
- Quaternary/Tertiary sedimentary rocks
- Upper Tertiary sedimentary rocks
- Lower Tertiary sedimentary rocks
- Tertiary volcanic rocks
- Franciscan Complex
- Great Valley Sequence
- Serpentinite, other Coast Range Ophiolite

ern part is in San Joaquin and Stanislaus Counties. Franciscan and Great Valley Complex bedrock are both present and complexly faulted together beneath the Santa Clara Valley.

Franciscan Complex

The parks in Santa Clara County are excellent places to see outcrops of many Franciscan rock types. They are exposed along the San Andreas Fault Zone in Los Trancos Open Space Preserve, Upper Stevens Creek Park, and Sanborn County Park; in Almaden Quicksilver County Park; and in the Diablo Range at Henry Coe State Park and Joseph Grant County Park.

Pillow basalt is common in the New Almaden Block in the foothills of the Santa Cruz Mountains, near the Almaden Reservoir, and on both sides of Uvas Reservoir, where the road on the west cuts through pillow basalt for a mile or so. Radiolarian chert can be seen on the road to Mount Umunhum and scattered through the mélange in the New Almaden area. In many places it occurs in tight folds like those at the classic Marin Headlands sites (pls. 17, 44). The Mount Hamilton Block in the Diablo Range is mostly graywacke and mélange, with some extensive outcrops of radiolarian chert. Most of the graywacke, like that in road cuts along Mount Hamilton Road to Lick Observatory, is somewhat metamorphosed.

Many beautiful high-grade Franciscan metamorphic rocks are scattered in mélange throughout the Diablo Range. Good examples can be seen along San Antonio Valley Road and along Kincaid Road near Mount Hamilton. Especially large metamorphic blocks occur at the southern end of Henry Coe State Park. If you enjoy hiking, Henry Coe is one of the best places in the South Bay to see Franciscan rocks. The entrance station is in mélange and has good exposures of the reddish radiolarian chert.

Four Franciscan terranes and mélange are represented in the South Bay (table 4): the Yolla Bolly and Burnt Hills Terranes in the Mount Hamilton Block; the Permanente and Marin Headlands Terranes in the New Almaden Block. You can see the "melted ice cream" topography characteristic of Franciscan mélange around Calero Reservoir, on Mount Hamilton Road, and throughout the Diablo Range. Go in the early morning or late afternoon, when the sun is low in the sky and the shadows emphasize the hummocky surface. The landscape looks as if it is flowing before your eyes. Throughout the Bay Area, erosion of mélange is one of the major processes forming the topography. The resistant rocks embedded in the South Bay mélange include somewhat metamorphosed graywacke, greenstone, chert, and the spectacular blueschist metamorphic rocks.

It may surprise you that rocks of the Marin Headlands Terrane are found 50 miles south of the Marin Headlands. The terrane was named for the excellent exposures in Marin County, but this combination of rocks, which includes chert, pillow basalt, and graywacke, is found at a number of Bay Area sites. This wide distribution probably reflects just how extensive the Marin Headlands Terrane originally was.

Great Valley Complex

Great Valley Complex rocks form a belt north and south of Loma Prieta in the Sierra Azul Block, and another long belt on the east in the Coyote and Alum Rock Blocks in the Diablo Range foothills (maps 19, 20). The latter belt extends north almost to San Pablo Bay. These Great Valley rocks include both the ocean crust material of the Coast Range Ophiolite and the overlying sedimentary rocks of the Great Valley Sequence (see chapter 3). The ophiolite is particularly well exposed along Hwy. 130 in Del Puerto Canyon (pl. 89) in the eastern Diablo Range, where Del Puerto Creek has cut through these interesting rocks that were brought up from the uppermost part of the earth's mantle during the Mesozoic plate collision. This is one of the best places near the Bay Area to see ophiolitic rocks.

Serpentinite, the California state rock and part of the ophiolite, is common throughout the South Bay, but especially so in the Silver Creek Block. The thin reddish soil and distinctive barren, rubbly slopes that characterize serpentinite areas (pl. 24) are very

Plate 89. Coast Range Ophiolite in Del Puerto Canyon. The white layer is sheared serpentinite.

Plate 90. Graywacke of the Great Valley Sequence behind bridge at Anderson Lake.

prominent in the hills east of Hwy. 101 where the Santa Clara Valley narrows and at Calero Reservoir. Good outcrops of serpentinite are exposed at the Chesbro and Almaden Reservoirs and at Santa Teresa County Park in the Santa Teresa Hills. The mercury deposits for which this area is famous are associated with altered serpentinite in the New Almaden area (see below), a good place to become familiar with this rock.

The marine sedimentary rocks of the Great Valley Sequence are common in the Sierra Azul and Coyote Blocks. The oldest of the sequence were deposited on the ophiolite. Much of the Coyote Block consists of graywacke from the youngest part of the Great Valley Sequence. These rocks can be seen at Coyote and Anderson Lakes, reservoirs that lie in the Calaveras Fault Zone. At Coyote Reservoir you can compare the Great Valley and Franciscan rocks: the ridge on the east side of the lake is Great Valley Sequence, and the ridge on the west side is Franciscan. At Anderson Lake, Great Valley graywacke is particularly well exposed at the east end of the bridge that crosses the south end of the reservoir (pl. 90). The blue green serpentinite on the west ridge makes a fine contrast with the gray brown graywacke of the east ridge.

The South Bay Tertiary Rocks: From Sea to Land

Many types of younger sedimentary rocks and interesting volcanic rocks were deposited on the Mesozoic Franciscan and Great Valley Complex rocks. After subduction ended, sandstone and shale began to accumulate in local marine basins that formed along the continental margin (see chapter 3). In the South Bay, Tertiary marine basin rocks are exposed in two belts, along the western edge of Santa Clara County in the Santa Cruz Mountains and on the east at the base of the Diablo Range. The rocks range in age from about 2 or 3 million to 50 million years. Lower Tertiary rocks that formed when the basins were still deep, about 50 to 25 million years ago, are present mainly in the western belt, a continuation of similar rocks in San Mateo County (see chapter 7). Along Skyline Blvd., which runs along the county line, you get occasional glimpses of marine sandstone and shale of the Butano and Vaqueros Formations, both common on the Peninsula. In many places, the Vaqueros Sandstone exhibits cavelike tafoni, which makes this rock a favorite of climbers in nearby Castle Rock State Park (pl. 80).

A long, almost continuous belt of upper Tertiary rocks, deposited as the marine basins became shallower, is present on the east side of the Santa Clara Valley. The rocks include the Claremont, Briones, and Orinda Formations that are common to the north in the East Bay Hills (see chapter 9). When the Claremont Formation was deposited, the marine basins were still many hundreds of feet deep, but the Briones Formation contains shallow marine fossils, and the Orinda Formation consists of sediments deposited by streams in a terrestrial environment. These rocks record the great change from ocean to land that took place toward the end of the Tertiary. You can see fine examples of several of these rocks in Alum Rock Park (see below).

Some outcrops of Tertiary sedimentary rocks are also present in the hills at the south end of the Santa Clara Valley. A small belt of lower Tertiary sandstone and shale lies along the northwest side of the Santa Teresa Hills. At Chitactac-Adams Heritage County Park, one of the upper Tertiary sandstones (the Temblor Sandstone) contains bedrock mortars made by Native Americans who lived along Uvas Creek (pl. 91). This small park, which preserves many cultural features, including petroglyphs, and has an

Plate 91. Temblor Sandstone with mortar at Chitactac-Adams Heritage County Park.

interpretive display of Ohlone culture, is a fine place to see both geology and history.

At the end of the Tertiary (1.8 million years ago), the Santa Cruz Mountains and the Diablo Range were rising and eroding. Streams carried sediment from the hills into a lowland valley, depositing sand and gravel on a broad alluvial plain. The alluvial deposits that came from the Diablo Range are called the Packwood and Irvington Gravels. Fossils found in them in the Irvington district of Fremont tell us of the animals that roamed the valley 1 or 2 million years ago. Picture a broad grassy alluvial plain with meandering streams and small lakes. Saber-toothed cats, camels, wolves, and mammoths wandered among the groves of willows, sycamores, and poplars that grew along the stream banks and in the oak woodlands along the hills at the base of the rising mountains. Fossils of many smaller vertebrates, including voles, gophers, mice, and pack rats, have also been found in these deposits.

The Santa Clara Formation, composed of stream deposits from the Santa Cruz Mountains, includes layers of conglomerate, sandstone, and shale that are poorly consolidated, that is, not firmly cemented into rock. You can get a fine look at them in Stevens Creek County Park on the road along the north side of the reservoir (pl. 92). The composition of the Santa Clara Forma-

Plate 92. Tilted layers of Santa Clara Formation at Stevens Creek Reservoir.

tion varies from place to place, reflecting uplift of different areas along the San Andreas Fault Zone and erosion of different types of rocks. A layer of ash in the upper part of the formation, the Rockland Ash, is about 570,000 years old. This is the same ash found in the sea cliffs at Daly City and in muddy sediments beneath San Francisco Bay. Tectonic activity in the past million or so years has tilted and folded these deposits, reminding us that fault activity in the South Bay continues to shape the landscape.

The youngest deposits in the South Bay are unconsolidated alluvial sediments deposited by streams in the Santa Clara and other valleys, and in lakes along the San Andreas and Calaveras

Faults, now mostly converted to reservoirs. Young mud is accumulating at the margins of San Francisco Bay (see chapter 6). These young deposits are the sedimentary rocks of the future.

Volcanic Rocks

Tertiary volcanic rocks are present in scattered areas of the South Bay (map 19). Some are related to the change from a converging to a sliding plate boundary; others, such as the volcanic rocks at Anderson and Coyote Reservoirs, are much younger, between 2 and 4 million years old. As elsewhere in the Bay Area, the volcanics are associated with faults, the fractures in the crust that permit magma to rise to the surface. In the South Bay volcanic rocks occur primarily along the San Andreas Fault, along the Calaveras Fault near Anderson and Coyote Reservoirs, and in the southern part of the Diablo Range. Volcanic eruptions about 15 to 16 million years ago produced the hydrothermal fluids that concentrated the mercury of the New Almaden Quicksilver District (see below).

Two volcanic units are worthy of mention although they lie primarily just south of the Bay Area: the 23-million-year-old Pinnacles in the Gabilan Range and the 9-to 11-million-year-old Quien Sabe Volcanics in the southern Diablo Range. The distinctive Pinnacles (pl. 93), south of Hollister in San Benito County, are especially interesting in rock type, in shape, and in their significance for movement along the San Andreas Fault. The rocks are remnants of a volcanic field that erupted in southern California along the San Andreas Fault in the Mojave Desert. The volcanics were split by the fault, and the Pacific Plate carried the western part 192 miles to northern California (fig. 13). We know that the two volcanics are related because they are highly unusual types of volcanic rock found nowhere else in the state except at these two localities. To add to their interest, they have eroded into strange shapes along prominent fractures in the rock to form a very unusual landscape. Explore this geological gem in spring when the weather is cool and the hillsides are covered with wildflowers.

The Quien Sabe Volcanics erupted as the San Andreas Fault System was becoming established in this area. They consist of a wide range of rock types, including rhyolite, dacite, andesite, and basalt. Because the rocks are resistant to erosion they form a

Plate 93. Volcanic rocks at the Pinnacles National Monument, San Benito County.

sharp and jagged topography on the eastern horizon as you drive along Hwy. 101, a strong contrast with the gentler landscape of the Franciscan rocks around them. They also provide important information about movement along the San Andreas Fault System. Recent research indicates that they are similar to the volcanic rocks at Burdell Mountain north of Novato that have been faulted northward more than 100 miles since they erupted about 10 to 12 million years ago.

Special Places to Explore

Like other parts of the Bay Area, the South Bay is rich in parks and open space that preserve the landscape and permit easy access to the botany, wildlife, and geology of the area. Many of the parks also have a significant story to tell of life in the South Bay when ranching and agriculture were the dominant activities. A visit to these parks rewards you with fine vistas of the varied landscape, from mountaintop to valley, and many opportunities to explore the geology at close range. Two areas of particular geological interest are described below.

Alum Rock Park: The South Bay's Hot Spring Resort

At Alum Rock Park in the Diablo Range foothills east of San Jose (map 18), you can enjoy a fascinating glimpse into the Bay Area's cultural past and hike on rocks that span more than 100 million years of Bay Area geologic history. This small park is one of California's oldest city parks and also one of the most interesting places to see some of the Tertiary marine sedimentary rocks common in the South and East Bay.

Go to the gate at the end of the road through the park to see fine exposures of the Claremont Formation (pl. 94), which at this site is composed of chert. It was formed about 14 to 16 million years ago in a deep marine basin. The chert occurs in thin layers that are beautifully folded and contorted, much like similar rocks in the East Bay Hills in Oakland. The layers were once flat and probably were tilted and folded as the Diablo Range was uplifted. Take a stroll along the Mineral Springs Trail to enjoy the grottos and baths built early in the twentieth century around springs that bubbled up along Penitencia Creek (pl. 95). Hot and cold mineral springs made Alum Rock one of the most popular parks in the

Plate 94. In Alum Rock Park, Claremont Formation layers that have been tightly folded.

Plate 95. Grotto and springs, Alum Rock Park. The white is a mineral deposit.

Bay Area in those days and a nationally known health spa. The springs are associated with the nearby Calaveras Fault. Farther east in Alum Rock canyon you can see sandstone of the Briones Formation, with layers of fossil mollusks like those at Mount Diablo (pl. 106).

Alum Rock itself (pl. 96), near the park entrance, and nearby Eagle Rock are volcanic. Recent studies have shown that they are of Mesozoic age and part of the Coast Range Ophiolite, not young volcanic rock (formerly called the Alum Rock Rhyolite), as described in previous works. Just west of Alum Rock near the park entrance are some outcrops of a conglomerate in the Berryessa Formation. This conglomerate is similar to and perhaps the same unit as the Novato Conglomerate in Marin County, another example of northward offset along one of the Bay Area's many faults.

Plate 96.
Alum Rock.

Almaden Quicksilver County Park

South of San Jose, in the foothills of the Santa Cruz Mountains, lies the site of one of the richest mining areas in California's history, the New Almaden Quicksilver District. Here, in the largest mercury mine in North America, mercury was mined from 1845 to the 1960s with small mining operations continuing until 1976. Santa Clara County purchased much of the property in 1973, and in 1975 the county opened the highly interesting Almaden Quicksilver County Park (map 18), where you can explore the geology and mining history of an area that played a major role in California's early days. Few of the mining structures remain today, but the history is well illustrated in the New Almaden Quicksilver Mining Museum.

New Almaden mercury mines were very important because mercury was an essential ingredient in the amalgamation process by which gold and silver were separated from their ores. The New Almaden Mine was opened just before the discovery of gold in

California, and it soon became the main source of the large amounts of mercury needed to recover gold. By the time the gold rush was over, the seven mines in the New Almaden District had produced more wealth than the gold mines themselves. Almost 3 million flasks of mercury, each weighing 76 pounds, were mined from the district, most from the New Almaden Mine, which was the largest and most lucrative mine in the district. The district has produced more mercury than any other mercury mine in the United States.

Plate 97. Los Capitancillos Ridge from Bald Mountain. Layered rock in left foreground is Temblor Sandstone.

Los Capitancillos Ridge (pl. 97) is one site of the former mines and today's park. The ridge is composed largely of blue green serpentinite, the state rock, with some pillow basalt blocks in Franciscan mélange. Altered serpentinite is the host rock of the mercury ore. The serpentinite is a soft rock with a greasy feel. Its shiny polished surfaces, called slickensides, formed when the serpentinite was squeezed and faulted under great pressure. It literally has polished itself. The rubbly serpentinite landscape is clothed in oak grasslands and chaparral, with fields of serpentinite-adapted wildflowers in spring.

Mercury, the only metal that is liquid at ordinary temperatures, is produced from the bright red ore cinnabar (mercury sul-

Plate 98. Hand sample of cinnabar, the ore of mercury. Sample is about 12 inches across.

fide) (pl. 98). Native Americans of the area used the soft cinnabar as a paint. The formation of cinnabar took place through several geologic steps. First, rock from the mantle was altered to serpentinite during Mesozoic subduction. The second step, about 15 to 16 million years ago, was the alteration of serpentinite to silica-carbonate rock, the host rock of cinnabar, by warm fluids associated with the change from a subducting to sliding plate boundary in this area. In the third step, hot fluids circulated through Great Valley Sequence sediments, picked up mercury (and other metals), and formed cinnabar. Finally, the cinnabar was deposited along fractures in the rock, setting the stage for future mining. You can see both serpentinite and silica-carbonate rock near the Day Adit in the park (pl. 99). (An adit is a horizontal tunnel into a mine.)

Mercury was first mined at New Almaden in 1845. Eventually many mines and more than 100 miles of tunnels were carved

Plate 99. Serpentinite (foreground) and silica carbonate rock (background) at Day Adit.

through the ridge (fig. 37). Today the main tunnel of the San Cristobal Mine is open for a short distance; here you can see the ore rock in the tunnel walls (take a strong flashlight). By 1865 there were hundreds of buildings and several thousand people lived nearby. Although most of the buildings have been removed, some of the old structures remain. A 70-foot-tall brick chimney looks out over the staging area, and parts of a rotary furnace built in the 1940s are left on Mine Hill. Particularly interesting is the Powder House, in which first black powder and then dynamite was stored for the mining operations. It was destroyed in the 1989 earthquake but has been rebuilt using the old bricks. The cinnabar was moved from the mine to the reduction furnaces by tramway, and the mercury product taken by horse cart to the port of Alviso for shipment by boat to the Sierra gold mines. The mining has left a legacy of mercury contamination of the Guadalupe River, which flows into San Francisco Bay. Recent studies have

Figure 37. Historic photo of Santa Isabel shaft by R. R. Bulmore or Dr. S. E. Winn, from *Views of New Almaden,* 1886.

shown that the New Almaden Mine is one of the major sources of mercury in the bay.

Cinnabar is found in other serpentinite areas in the Coast Ranges, but nowhere in such large deposits as these. The New Almaden Mine and the New Idria Mine to the south in San Benito County were the two most important mercury mines in California. Mercury mines were also established in the North Bay, in Napa, Sonoma, and Lake Counties, and small deposits occur on the east side of Mount Diablo. None of these mines are still working.

At Almaden Quicksilver County Park, 33 miles of hiking, biking, and horseback trails provide access to the former mining area. All the mines themselves are dangerous and have been sealed. The toxic tailings piles have also been removed. You can explore the history of mercury mining at New Almaden in the excellent mining museum, which shows photographs and artifacts from the mining days. The museum is housed in the historic La Casa Grande, which was the home of mine managers for 58 years.

CHAPTER 9
THE EAST BAY

A Young Landscape

DRIVING THE BACK ROADS through the peaceful hills and valleys of the East Bay (map 22), you would not guess how young and dynamic this landscape is. Even in the geologically complicated Bay Area, the East Bay stands out for its fault activity, diversity of rocks, and geologic variety (map 23). The hills of Alameda and Contra Costa Counties are at most a few million years old and are still going up. And the rock record preserves much of the Bay Area's dramatic geologic history. In particular, here we can see the changes from a colliding to a sliding plate boundary, from deep ocean to shallow sea and finally to land.

More than a dozen active faults slice through the East Bay, rearranging the landscape and the rocks. They are part of the San Andreas Fault System, the sliding boundary between the Pacific and North American Plates (see chapter 2). In the East Bay, movement between the plates is divided among many faults, unlike in Marin County, for example, where the dominant movement takes place along the San Andreas Fault. A component of compression between the plates is expressed by folding and faulting that has formed the young hills of the East Bay and thrust up Mount Diablo in the past million years or so.

The East Bay landscape reflects both the pattern of faults and that of rock types. Along San Francisco Bay lie the "flatlands," as they are called by locals. This highly urbanized area is a gently sloping alluvial plain formed of bits of rock eroded out of the rising East Bay Hills. Over the past million years the many creeks that drained the hills carried sediment toward the bay and built up the plain. Three major creeks, San Leandro, San Lorenzo, and Alameda, have formed large alluvial fans as they carried sediment toward the bay. Today, most of the creeks are culverted or buried beneath the cultural overlay, but if you follow the dips of neighborhood streets that parallel the hills, you can tell where they are. Here and there, you can still see bay or willow trees that grew along the creeks. Growing interest in creek restoration is helping to "daylight" some of them.

Just east of the alluvial plain, the long ridge of the East Bay Hills stretches from San Jose to San Pablo Bay. Rocks spanning the geologic history of the Bay Area from the Mesozoic plate collision to young stream deposits can be found in these hills. They are actively rising today, squeezed up between two major

Figure 38. Berkeley Hills without trees, looking southeast toward head-waters of Wildcat Canyon. Highest peak is Grizzly Peak.

faults, the Hayward Fault on the west and the Calaveras Fault on the east (map 24). The East Bay Hills of today look very different from those of 100 years ago, not just because of the houses and roads, but also because they are tree covered, mostly with introduced eucalyptus and Monterey pine. Photographs taken around 1900 show grassland hills with native redwood and bay trees in the moister canyons and oaks on the drier exposures (fig. 38).

Farther east, the landscape is one of wide valleys and low rolling hills, with Mount Diablo rising majestically above them in the north and the rugged Diablo Range in the south (map 22). Both of these high areas are islands of Mesozoic rock in a sea of younger sedimentary rock. But although the rocks are old, this landscape, too, is deceptively young. The low hills and Mount Diablo itself are rising today. The rolling hills east of Richmond are formed of relatively young (Tertiary) marine and terrestrial sedimentary rocks. They form a soft, sensuous landscape, like that of the Tassajara Hills, south of Mount Diablo.

The urbanized north-south valley from Concord to Pleasanton, the Hwy. 680 corridor, is underlain by alluvium washed out of the surrounding hills. The broad east-west Livermore Valley is a down-dropped basin between Mount Diablo and the Diablo Range that is filled with young sediments. East of the rolling hills lie the flatlands and meandering sloughs of the Delta, a bit of Central Valley geology that is shared with the Bay Area.

text continues on page 224

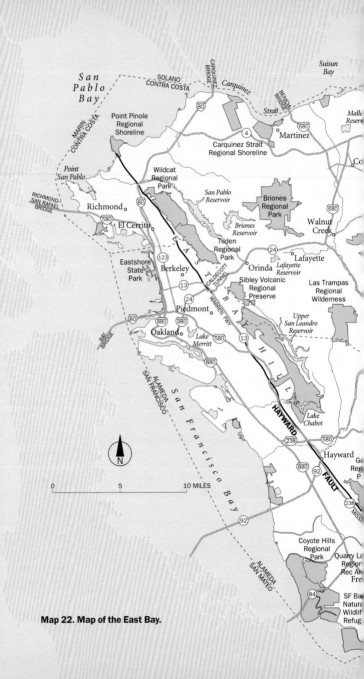

Map 22. Map of the East Bay.

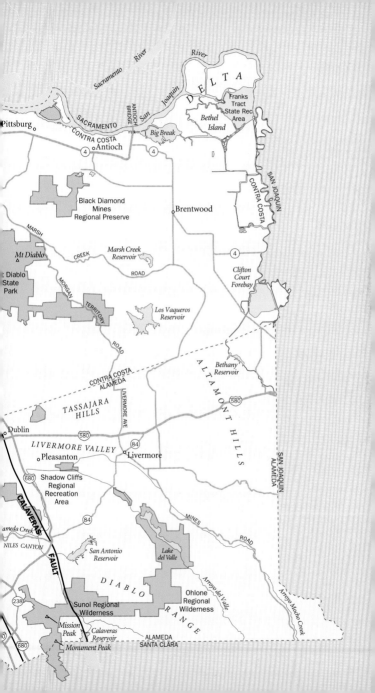

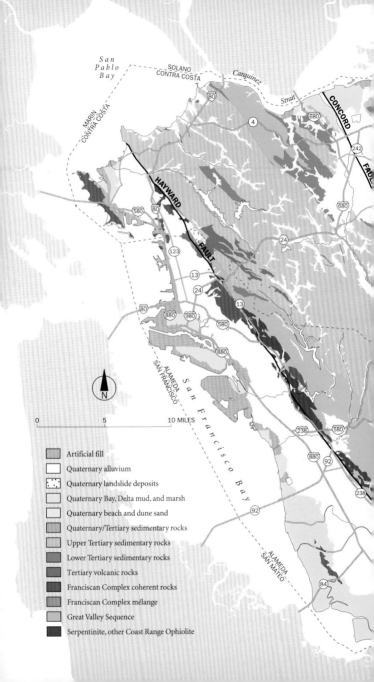

San Pablo Bay

SOLANO
CONTRA COSTA

Carquinez

Strait

CONCORD

80

4

680

242

FAULT

MARIN
CONTRA COSTA

HAYWARD

580

80

680

123

FAULT

24

13

24

13

80

880

980

580

880

ALAMEDA
SAN FRANCISCO

N

San Francisco Bay

238

580

880

92

0   5   10 MILES

92

238

ALAMEDA
SAN MATEO

84

Artificial fill

Quaternary alluvium

Quaternary landslide deposits

Quaternary Bay, Delta mud, and marsh

Quaternary beach and dune sand

Quaternary/Tertiary sedimentary rocks

Upper Tertiary sedimentary rocks

Lower Tertiary sedimentary rocks

Tertiary volcanic rocks

Franciscan Complex coherent rocks

Franciscan Complex mélange

Great Valley Sequence

Serpentinite, other Coast Range Ophiolite

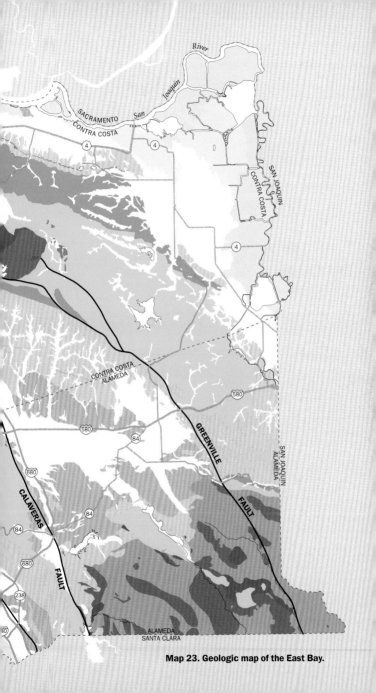

Map 23. Geologic map of the East Bay.

The Delta: A Subsided Marsh

Before development, vast wetlands lined the Sacramento and San Joaquin Rivers. Extensive marshes and a network of sloughs existed where the rivers joined. Each spring the rivers flooded, bringing fresh sediment and nutrients to the marshes. As marsh vegetation died, it accumulated into thick peat deposits that built up on the islands. In the late 1800s levees were built around the Delta islands and the rich peat soil began to be farmed. Today fields of feathery asparagus wave in the Delta breezes in one of California's richest agricultural areas.

When the levees were built, the annual sediment input was cut off, and the peat soils compacted, dried out, and began to blow away. Behind the levees the island surface got lower and lower. Today you can drive on levee roads and look out over the tops of pear trees or at fields 10 to 15 feet below you. Imagine what would happen if the levee broke. In the northeast corner of Contra Costa County you can see the result. Drive to the eastern edge of Bethel Island and gaze east at Franks Tract State Recreation Area (map 22), a former island that flooded in 1936 when a levee broke (pl. 100). Now it is a large expanse of water with a great number of birds and healthy wetland vegetation. Traces of the former levee system remain above water. Big Break, just east of Antioch, is another such flooded island. Levees on other islands have broken over the years, but if the agriculture on the island is very valuable, it is worth the economic cost to fix the levee and pump the water out.

Creeks That Go the Long Way Round

The creeks that drain the East Bay all flow into San Francisco Bay, but some take a very circuitous path (map 24). West of the East Bay Hills, creeks flow directly across the alluvial plain into the bay, although several rest at reservoirs on their way. Creeks east of the hills take a much longer route. San Pablo and Wildcat Creeks flow north along the hills and into San Pablo Bay. Walnut Creek flows northward into the wetlands of Pacheco Creek and into Suisun Bay. Alameda Creek flows southward along the hills, cuts its way through them at Niles Canyon (Hwy. 84), then goes to the bay. All the creeks east of Mount Diablo flow to the Delta, which drains to the bay via San Joaquin River tributaries.

Plate 100. Aerial view of Franks Tract in 1978. Note remnants of former levees.

Alameda Creek is quite unusual among Bay Area streams. You would expect it to flow around the East Bay Hills as Walnut Creek does, rather than through them. But Alameda Creek is an antecedent stream, one that is older than the hills through which it flows. It was flowing westward from the Livermore Valley to the bay before the hills started going up about 1 million years ago. As the hills began to rise, erosion kept pace with uplift and the creek carved Niles Canyon. Alameda Creek is the longest and largest creek in the East Bay. It drains 700 square miles, collecting many subsidiary creeks on its way to the bay. The sediments carried westward by Alameda Creek built up a large alluvial fan along the edge of the bay. The sand and gravel that make up the fan are excellent aquifers and have long served as a source of drinking water for southern Alameda County. Where Alameda Creek crosses the Hayward Fault, it is offset about 1,000 feet to the north by fault movement. Today its waters are diverted into abandoned quarry pits along the fault at Quarry Lakes Regional Recreation Area. The pits are used as percolation ponds to recharge the underlying aquifer, called the Niles Cone.

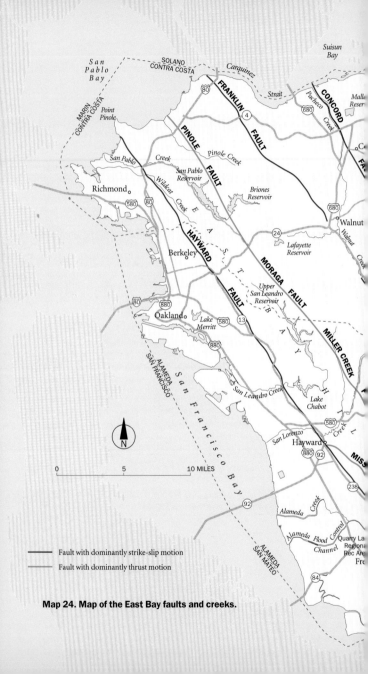

Map 24. Map of the East Bay faults and creeks.

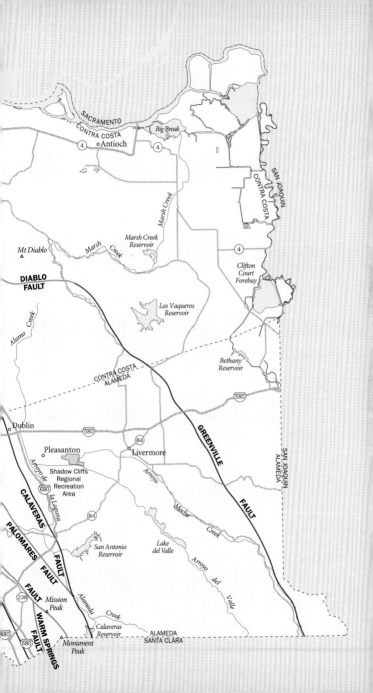

The East Bay Fault System

In the tectonically active East Bay, the landscape is being re-arranged rapidly, geologically speaking, along the many faults that cross the area (map 24). We are all familiar with the Hayward Fault which runs northward along the west front of the East Bay Hills to San Pablo Bay at Point Pinole Regional Shoreline (pl. 101). The Calaveras, Concord, and numerous other faults more or less parallel it to the east. Each of these also has subsidiary strands, creating a complex network of faults known as the East Bay Fault System. The dominant movement on these faults is right-lateral strike-slip, like that on the entire San Andreas Fault System (see chapter 2). The complexity of East Bay geology is due to this maze of faults, which divide the East Bay into more than a

Plate 101. Aerial view of Hayward Fault looking north to Carquinez Strait. Hwy. 13 (Warren Expressway) in foreground; Hwy. 24 across center.

dozen slivers of rock, each with its own geologic story. Movement on the East Bay Fault System shifts each sliver northwestward relative to the one east of it (fig. 16), in the direction that the Pacific Plate is sliding past the North American Plate. Over time this has thoroughly rearranged the rocks and juxtaposed very different types of rocks.

The East Bay Fault System was initiated about 12 million years ago as movement along the Hayward Fault began in this area. The East Bay faults have not all been active at the same time. Movement may shift from one fault to another and sometimes shift back. Over the past 12 million years, about 110 miles of offset has occurred along the East Bay faults, mostly on the Hayward Fault, moving rocks that originated to the south northward to their present location. The highest part of the East Bay Hills, from Mission Peak to Monument Peak in Fremont, occurs where the Hayward and Calaveras Faults almost converge, and movement transfers, or "steps over," from one fault to another (see chapter 2).

In addition to strike-slip faults, there are many thrust faults in the East Bay. They shove older rocks over younger ones, as the Mount Diablo Fault is uplifting Mount Diablo. These young and active faults are part of the transpressional processes that have affected the Bay Area in the past 3.5 million years or so.

The Hayward Fault

The most prominent of all the East Bay faults is the Hayward Fault (map 24). For much of its length it is marked by the typical features of a fault zone (fig. 12): landslides, springs, offset streams, and linear valleys like the one through which the Warren Freeway (Hwy. 13) runs. Several East Bay communities were originally founded along what we now know to be the Hayward Fault because of plentiful water from fault-related springs. The crushed rock along the fault acts as a barrier to subsurface water draining from the hills, and this water surfaces as springs along the fault. Mission Blvd. (Hwy. 238) between Hayward and Fremont has several old buildings, such as the Masonic Home and Mission San Jose, built to take advantage of the springs.

The last major earthquake on the Hayward Fault was in 1868 (magnitude about 7). The ground ruptured for a distance of 20 to 30 miles and was offset up to 6 feet horizontally. Every building in Hayward was damaged, many demolished, and about 30 people died. Historic reports of an 1836 earthquake on the northern segment of the Hayward Fault were recently shown to be incorrect. Evidence from trenches dug across the fault at the Mira Vista Golf Course east of Richmond indicates that the last rupture of that segment was between 1640 and 1776. Because the northern Hayward Fault has not moved in a long time, the probability of an earthquake may be higher for this segment than elsewhere.

The Hayward Fault also moves by fault creep along almost its entire length. Creep is a slow movement that does not produce earthquakes or shaking. Creep meters in Berkeley under the University of California Memorial Stadium have recorded intermittent movement there averaging about one- to two-tenths of an inch per year. Offset curbs in Hayward and bent railroad tracks in Fremont are visible reminders of creep along the fault (pl. 102).

Plate 102. Curb offset at Rose and Prospect Streets, Hayward, 1989.

Rocks Old and Young

The rolling grassland and oak hills that make up much of the East Bay landscape cover an amazing variety of rock types (map 23), from rocks that formed during Mesozoic subduction more than 100 million years ago to young unconsolidated sedimentary deposits. They record a long geologic history, including a major change — the transition from ocean to land. Because of the many active faults that have carved up and rearranged the East Bay, we have to piece the geologic story together from several different localities.

The Mesozoic Subduction Record

The basement, or oldest, rocks in the East Bay are the same Franciscan and Great Valley Complex rocks found elsewhere in the Bay Area, rocks that were formed during more than 100 million years of Mesozoic subduction (see chapters 2 and 3). They include both accreted terranes that have come from great distances and rocks formed regionally during subduction. Throughout the East Bay, the Franciscan and Great Valley basement rocks have been faulted together by ancient subduction-related faults, as well as by the present East Bay Fault System.

Franciscan Complex

All the Franciscan Complex rock types — basalt, chert, graywacke, and metamorphic rocks — can be seen in the East Bay, particularly in the Diablo Range and on Mount Diablo. Although Franciscan bedrock underlies most of the Bay Area east of the San Andreas Fault, only a few Franciscan outcrops stick up through the alluvial plain in the East Bay (map 23). Four of the Franciscan terranes and mélange are represented (see chapter 3 for a description of the different rocks and terranes). Good exposures of the graywacke and shale layers of the Novato Quarry Terrane can be seen at the east end of the Richmond–San Rafael Bridge and at Point San Pablo (pl. 103), in the Eastshore State Park, at nearby Albany Hill, and in Piedmont.

The beautiful reddish radiolarian chert of the Marin Headlands Terrane is particularly well exposed at Coyote Hills Regional Park in Fremont (pl. 104). There, a climb to the top of Red Hill also gives you an expansive view of the bay to the west and marshes to the east. Franciscan basalt has been mined for construction material in a large quarry at the southern end of the Coyote Hills, just south of the Dumbarton Bridge toll plaza. A pedestrian overpass provides a link from the park trails to trails around the quarry rim. A variety of Franciscan rocks of the Marin Headlands Terrane and mélange can be seen at Mount Diablo (see below). Metamorphic rocks embedded in Francis-

Plate 103.
Franciscan
graywacke and
shale of the
Novato Quarry
Terrane at
Point San Pablo,
Richmond.

Plate 104.
Franciscan chert
in Coyote Hills
Regional Park.

can mélange occur in El Cerrito, in Kensington, and in Little Yosemite in Sunol Regional Wilderness (see chapter 3, opening photo). The northern end of the Diablo Range south of Livermore has Yolla Bolly Terrane graywacke, which is also present in the East Bay Hills in El Cerrito. A few exposures of Alcatraz Terrane graywacke occur in the East Bay Hills, for example, in Kensington, just south of the Sunset View Cemetery.

Great Valley Complex

Large outcrops of the Coast Range Ophiolite and the overlying Great Valley Sequence of marine sedimentary rocks (map 23) occur in the East Bay. Discontinuous outcrops of serpentinite, part of the ophiolite, can be seen along the Hayward Fault from Richmond south to Fremont, in the Diablo Range, and on Mount Diablo (see below). An unusual volcanic rock in the East Bay Hills at Oakland, the Leona Rhyolite, is also part of the ophiolite. This rock was quarried in the huge Gallagher & Burk quarry along Hwy. 580, whose scar is easily visible from across the bay. Pyrite (iron sulfide), chalcopyrite (iron copper sulfide), sulfur, and, later, fill for the Port of Oakland were quarried here.

A long belt of the marine sedimentary rocks of the Great Valley Sequence is present in the East Bay Hills east of the Hayward Fault from Berkeley southeast to Santa Clara County. Road cuts along Skyline Blvd., Pinehurst Road, and Redwood Road in Oakland give you glimpses of sandstone, shale, and the unusual Oakland Conglomerate. It contains large volcanic cobbles and other interesting rocks whose place of origin is not known but is likely to have been hundreds of miles to the south. The exposures along Skyline Blvd. for several miles south of Redwood Road are an excellent place to see this rock.

A small belt of Great Valley rocks lies south of Carquinez Strait in the hills west of Martinez. Good outcrops of sandstone can be seen along trails of Carquinez Strait Regional Shoreline. The largest mass of Great Valley rocks extends from Mount Diablo south to the Altamont Hills. Marsh Creek Road follows a valley cut into a soft shale, and road cuts give you a chance to examine some of the Great Valley rocks more closely. They were deposited by turbidity currents (see chapter 3). Good examples of sandstone-rich layers deposited closer to shore can be seen at a turnout on the north side of the road, 3.2 miles east of Morgan

Plate 105. *Top,* Great Valley sandstone (light brown) and shale (thin dark layers) along Marsh Creek Road. *Bottom,* Fine-grained turbidites, near Altamont Pass.

Territory Road (pl. 105, *top*). Fine-grained shale layers deposited farther from shore are present near Altamont Pass (pl. 105, *bottom*). All the East Bay Great Valley Sequence rocks are assigned to the Del Puerto Terrane.

From Sea to Land

Deposited on the Franciscan and Great Valley Complex basement rocks in the East Bay is a thick sequence of younger sedimentary and volcanic rocks (map 23) that provide important clues to the geologic history of the Bay Area after subduction ended. They tell the story of how this area emerged from the sea and became land. As the San Andreas Fault System began in southern California about 25 million years ago, faulting formed more restricted regional basins in areas that had been deep ocean (see chapter 2). Rocks of that age indicate that water depth in these basins was sometimes deep and at other times shallow because of sea level changes or movement of the crust.

Microfossils in the rocks show that the long range trend was to shallowing water and gradual uplift. Rocks formed in the East Bay by about 11 to 12 million years ago contain fossil mollusks that lived in shallow water near shore. They are preserved in rocks called the Briones Formation (pl. 106), one of the most interesting and widespread rock units in the East Bay (pl. 107). It is a resistant sandstone that forms prominent ridges from Briones Regional Park to Mission Peak.

Plate 106. Fossil clams and wave-rounded pebbles in the Briones Formation on Shell Ridge, Walnut Creek.

Plate 107. Ridge of Briones Formation at Las Trampas Regional Wilderness looking north. Bollinger Canyon in the foreground and Mount Diablo in the background. The valley is eroded into fine-grained sedimentary rocks.

Somewhat younger rocks are no longer marine; they are terrestrial deposits, indicating that by about 10 million years ago this area was uplifted above sea level. Fossils found in Livermore Valley deposits and on the lower slopes of Mount Diablo give us a fine picture of what the landscape was like. A very rich fossil site at Blackhawk Ranch on the south side of Mount Diablo contained bones of saber-toothed cats, horses, mastodons, camels, and small mammals that lived in the area about 9 to 10 million years ago (fig. 39). Similar fossil remains were found near Orinda and in the San Joaquin Valley. Younger sediments, between 10,000 and 2 million years old, contain different vertebrate fossils, including mammoths, ground sloths, huge bison, and many other animals. They roamed a hilly landscape more like that of today. From the many fossils found in the East Bay we know that for more than 10 million years large mammals and many other animals roamed a rich grassland that stretched across the East Bay and into the Central Valley.

Figure 39. Miocene vertebrates living in the Livermore Valley.

The Livermore Valley

A good example of the processes of erosion, deposition, and tectonics that shaped the East Bay terrestrial landscape is the Livermore Valley. It has been a lowland basin for millions of years, and parts of the valley are filled with more than a mile of gravel and sand, deposited by streams flowing into this basin from all sides. The younger sands and gravels deposited in the past 2 million years or so are called the Livermore Gravels in the southern part of the valley and the Sycamore Formation (or Tassajara–Green Valley Formation) in the lovely rolling Tassajara Hills to the north.

The gravel and sand of these formations are poorly consolidated; that is, they have not been firmly cemented or compressed together and are not quite rock. You can see what they look like in former quarries at Shadow Cliffs Regional Recreation Area (pl. 108), where the Livermore Gravels were quarried for construction material. The alternating sand and gravel layers reflect the size of the material that streams can carry at any particular time. Streams carry coarser materials like gravels during floods, and finer sands and silts when they flow gently.

Studies of cores taken from older basin sediments beneath the valley tell us that the composition of the basin fill has changed over time. From these deposits we can read the story of how the local landscape developed. Before today's Livermore Valley

Plate 108. Livermore Gravels at Shadow Cliffs Regional Recreation Area, Livermore Valley.

Plate 109. Flat river terrace (center) in Livermore Gravels along Mines Road in Arroyo Mocho, looking north toward Livermore. Mount Diablo in left background.

formed, rivers from the Sierra brought sediment to the area, including rocks eroded from active volcanoes erupting to the east. Streams also carried sediment from the Diablo Range into the valley from the south. Mount Diablo, the East Bay Hills, and the Altamont Hills did not yet exist. About 6 million years ago the Altamont Hills began to rise, cutting off the Sierra rivers and depositing locally derived sediment on top of the older gravels and sands. And finally, in the past few million years, Mount Diablo and the East Bay Hills began to go up and their streams added debris to the valley fill.

Although the sediments beneath the valley are young, they are folded into a gentle syncline, a reminder that tectonic movements are still going on in the area. You can see interesting evidence of this uplift as you drive southward on Mines Road out of Livermore. As you climb into the hills along Arroyo Mocho Creek, outcrops of uplifted Livermore Gravels and former river terraces line both sides of the road (pl. 109), a visible reminder of the East Bay's ongoing tectonic activity.

Special Places to Explore

Mount Diablo State Park

Mount Diablo is a magical place—a small mountain of ancient rocks (map 25) with an extraordinary view of California from the top. Although it looks a little like one, Mount Diablo is not a volcano. The volcanic rocks near the top are from volcanoes that erupted long ago far to the west and deep beneath the ocean. Mount Diablo's outward resemblance to a volcano comes from the same geologic processes that create the rest of our landscape—tectonic uplift and erosion. Although Mount Diablo rocks are quite old, the mountain itself is very young, geologically speaking. The evidence includes deposits from a 4-million-year-old volcanic tuff that originated in the Sonoma Volcanics in the North Bay (see chapter 10). The tuff is present just south of Mount Diablo and must have flowed across a relatively flat landscape. Neither Mount Diablo nor Carquinez Strait could have existed at the time because they would have interrupted the flow.

Mount Diablo stands above the landscape today because it is still being thrust up by tectonic forces and because the Franciscan and Great Valley rocks at the top are more resistant to erosion than the softer and younger sedimentary rocks that surround the mountain. Mount Diablo is a product of the slight compression

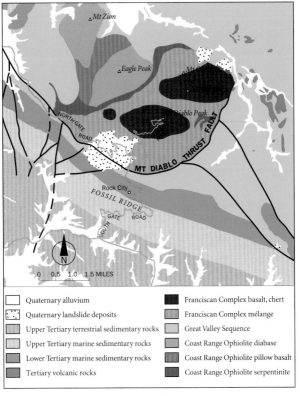

Map 25. Geologic map of Mount Diablo State Park.

Legend:

- Quaternary alluvium
- Quaternary landslide deposits
- Upper Tertiary terrestrial sedimentary rocks
- Upper Tertiary marine sedimentary rocks
- Lower Tertiary marine sedimentary rocks
- Tertiary volcanic rocks
- Franciscan Complex basalt, chert
- Franciscan Complex mélange
- Great Valley Sequence
- Coast Range Ophiolite diabase
- Coast Range Ophiolite pillow basalt
- Coast Range Ophiolite serpentinite

between the Pacific and North American Plates that is folding and faulting all of the Bay Area (see chapter 2). In past studies, the uplift of Mount Diablo has been ascribed to other processes. It has been called a diapir or a piercement, in which the older rocks now at the top pushed up through the younger sedimentary rocks. With today's better understanding of tectonic forces, the uplift is explained as the result of movement on the Mount Diablo Thrust Fault (map 25).

From the top of Mount Diablo on a very clear day you can see a remarkable expanse of California (pl. 110)—from the Farallon

Plate 110. Winter view from top of Mount Diablo to northeast across the Delta and Central Valley to the snow-covered Sierra Nevada, west of Lake Tahoe.

Islands, 65 miles to the west, to Yosemite on the east, and perhaps even Half Dome on a particularly clear day. You can look northeast to Lassen Peak about 180 miles away and southeast to the high Sierra crest. This exceptionally unobstructed view is possible because Mount Diablo is surrounded only by low hills and valleys. Winter mornings after a storm are the best; that is when the air is sparkling clear. At sunset another extraordinary sight greets you from the top; you can watch Mount Diablo's triangular shadow spread eastward into the Central Valley.

The rocks at the top of Mount Diablo are as interesting as the view. They date from the time of Mesozoic subduction (see chapters 2 and 3). Franciscan rocks underlie Mount Diablo Peak, North Peak, and Mount Olympia on the northeast side of the mountain. Fine examples of tightly folded reddish chert and pillow basalt are exposed along Summit Road. In between the rocky outcrops, Franciscan mélange forms the landscape. The

Plate 111. Uptilted Great Valley Sequence sandstone ridges, looking toward Antioch.

Franciscan rocks on Mount Diablo are like those at the Marin Headlands and belong to the Marin Headlands Terrane.

On Mount Diablo's northwestern side, Eagle Peak and Mount Zion are underlain by Great Valley Complex rocks, including basalt and diabase of the Coast Range Ophiolite, the ocean crust and mantle rocks faulted up during the Mesozoic subduction. A thin belt of serpentinite, our state rock, with its distinctive blue green color, separates Great Valley from Franciscan Complex rocks. At lower elevations lie Great Valley Sequence marine sedimentary rocks, originally deposited onto the ophiolite. They form the rocky ridges on the eastern and southeastern flanks of the mountain that were pushed up as Mount Diablo was thrust up. Each ridge is the edge of a resistant sandstone; the valleys between them are formed of less resistant shale layers. These tilted-up layers form a dramatic sight from the top (pl. 111).

The lower part of Mount Diablo on the west and southwest consists primarily of Tertiary marine and terrestrial sedimentary rocks, mainly sandstone and shale. They include layers of the fossiliferous Briones Formation (pls. 106, 107). The observation tower was constructed of this rock. Fossil clams, oysters, scallops, and mussels can be found in the blocks as you climb the stairs to the observation platform. In some places the rock layers have been tilted up and now are almost vertical (pl. 112). The sand-

Plate 112. Briones Formation at Blackhawk Ridge, just south of South Gate entrance station, looking east. Note change in vegetation between chaparral on sunny south-facing slope and trees on shady north slope. The same rocks continue across the road on Fossil Ridge.

stone Wind Caves at nearby Rock City (map 25) are among many other interesting geologic features to investigate at Mount Diablo. The caves are formed not by wind but by rainwater dissolving the cement that holds the sand grains together (pl. 113). To see many of these rocks and get a sense of Mount Diablo's long geologic history, take the park's 8-mile hike on the Trail Through Time. Starting at the base of the mountain in 9-million-year-old stream deposits at Blackhawk Ranch, you wind up through older and older rocks of the Tertiary Period. Then you cross a fault into Cretaceous rocks dating from the time of the Mesozoic subduction and finally end in 190-million-year-old ocean crust near the top. It is a steep climb, but where else can you walk back through almost 200 million years in just a few hours? Exploring the trails of Mount Diablo State Park rewards you with marvelous vistas and highly interesting geology. Also explore the exhibits in the visitor's center at the top to learn more about this fine park.

Plate 113. Caves in sandstone at Rock City, Mount Diablo State Park.

Black Diamond Mines Regional Preserve

In Black Diamond Mines Regional Preserve, the rocks tell the geologic story of the East Bay after subduction ended. Here you can see another record of the change from deep ocean to land, visit a sand mine, and learn the story of coal mining on the slopes of Mount Diablo.

Black Diamond Mines Preserve is named for the coal deposits, the "black diamond," that formed as the land in this area began to emerge from the sea. Coal forms in lagoons and swamps of coastal plains in subtropical climates. The presence of coal in northern Contra Costa County tells us that shoreline conditions existed at the time the coal formed, about 40 million years ago, and that the climate was much warmer than it is today. More than 4 million tons of coal were mined here from the 1860s to about 1900. The Black Diamond field is the largest known coal deposit in California. For decades it supplied the fuel that kept San Francisco warm and fired the furnaces of mercury mines and industry. The coal seams occur as layers in estuarine sandstone of the Domengine Formation, much of which is a beautiful white sandstone you can see along trails through the preserve (pl. 114). At

Plate 114. Domengine Formation sandstone at Black Diamond Mines Regional Preserve.

times when sea level was low or tectonic movements uplifted the land, layers of coal formed in swamps along the shore; then as sea level rose or the land subsided, sand was deposited on the coal layers. The sand at Black Diamond is a high-grade silica sand, composed mostly of quartz grains and clay flakes, the perfect combination for glass making. It was mined from the 1920s to the 1940s.

Today, after a great deal of work to make the Black Diamond area safe, you can visit the underground workings of the sand mine. Guided tours take you through some of the tunnels and into the huge "rooms" where sand was mined. The coal mines below are not open to the public because toxic gases accumulate in them.

In addition to the mine, the preserve is an especially good place to see the relationship between rock type and landscape. The rock layers have been tilted on their side, with the oldest rocks toward Mount Diablo. The Domengine Formation is very resistant to erosion and underlies the prominent ridge above the mines (pl. 115). The next-younger rock layer is the softer Nortonville Shale, which forms the valley just north of the mines. The mining town of Somersville was in this valley, and the Rose Hill Cemetery is on the slopes above it, offering poignant testimony

Plate 115. Black Diamond Mine area in 1977. Domengine Formation sandstone underlies tree-covered ridge on right; Nortonville Shale forms the valley; Markley Formation sandstone is prominent grassy ridge on left. Note mine dumps, now removed. Rose Hill Cemetery (foreground) has now been restored.

to the rigors of coal mining and life in a mining town. Fossils in the Nortonville Shale show that it was deposited in a deep marine basin, further evidence of changes in sea level or tectonic movements of the crust. On the north side of this valley is another ridge of resistant sandstone, the Markley Formation, with prominent caves eroded into its face.

The road from Antioch to Black Diamond Mines Regional Preserve goes through Markley Canyon, a canyon cut across the ridges of time. The tilted rock layers get older as you drive into the park. The preserve headquarters sits on Sydney Flat Shale, an easily eroded rock between two resistant sandstone layers of the Markley Formation. As you drive out of the preserve, the road cuts through ridges of increasingly younger and shallower marine sandstones that contain fossil mollusks indicative of nearshore and estuarine environments. The most northerly ridge is rock with a bluish cast, the Neroly Formation, which is about 9 to 11 million years old. It consists of volcanic sediments deposited by a river system flowing west from the Sierra Nevada, and it tells us that by this time, northern Contra Costa County was land.

Thus, on a visit to Black Diamond Mines Regional Preserve you can read the geologic story of the East Bay from 10 to 40 million years ago.

The East Bay Hills

The sedimentary rocks in the ridge of hills that extends from San Pablo Bay south to the Diablo Range also records the change from deep marine to terrestrial environments, but here different rocks provide the clues. Spectacular road cuts along Hwy. 24 just east of the Caldecott Tunnel between Oakland and Orinda (pl. 116) are an excellent place to see the marine to terrestrial change and to wonder at how the rocks have been folded. If you are stuck in slow traffic, there is much interesting geology to take your mind off the traffic. Here, a sequence of layered rocks representing marine, alluvial, and volcanic environments are exposed along the highway. These rocks were more or less flat lying when they formed, 9 to 16 million years ago. But as the East Bay Hills were pushed up, the rock layers were folded and faulted to their present positions. The section between the Caldecott Tunnel and the Orinda exit is folded into a broad syncline (down fold) (fig. 40). Erosion-resistant volcanic rocks form the high ridges at

Plate 116. Sandstone and conglomerate of Orinda Formation (foreground) and Moraga volcanic rocks (right), east of the Caldecott Tunnel.

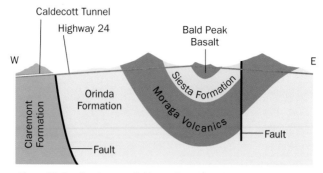

Figure 40. Syncline between Caldecott Tunnel and Orinda exit on Hwy. 24.

either end of the syncline, and soft, easily eroded lake-bed sediments are exposed in the middle at Gateway Blvd.

The Caldecott Tunnel was blasted through a ridge of the resistant Claremont Formation (pl. 117). This rock is about 14 to 16 million years old and was formed in a deep marine basin. It is made up largely of the skeletons of microscopic single-celled plants and animals called diatoms and foraminifers. Much of the Claremont Formation consists of thin bands of whitish chert, well exposed along Skyline Blvd. above the tunnel and to the south. This rock is similar in type and age to the Monterey Formation found at Point Reyes and on the Peninsula, but it was deposited in a different basin. The Claremont Formation tells us that about 14 million years ago the area in which it formed was still marine.

If you unfold the hills and flatten them out, the rocks lying on top of the Claremont Formation are younger layers of sandstone and conglomerate called the Orinda Formation. In the Hwy. 24 road cut, they are the distinctive red and gray layers on both sides of the freeway as you leave the east end of the Caldecott Tunnel (pl. 116). These rocks, which are 10 to 12 million years old, are not marine; they were deposited by streams flowing across an alluvial plain, much like the sloping plains surrounding San Francisco Bay today. About 2 million years of rock are missing between the Claremont and Orinda Formations, probably removed by fault action. During the time represented by this missing rock record, the area was uplifted and became land.

The Orinda Formation has another interesting story to tell.

Plate 117. Claremont Formation on Claremont Avenue, Berkeley. The layers are steeply tilted.

Some pebbles in the conglomerate are distinctive Franciscan rocks, and their arrangement shows that the streams that deposited them flowed from the west. The Orinda pebbles were deposited before the East Bay Hills were uplifted, when the land to the west across the Hayward Fault was a ridge of Franciscan rocks, not a valley (filled today by San Francisco Bay). Creeks draining the ridge carried sediment eastward to the edge of the shallow marine sea, where the towns of Orinda, Lafayette, and Walnut Creek are now. In the 10 to 12 million years since the Orinda Formation was formed, the Franciscan rocks from which the pebbles came have moved northward about 60 miles along the Hayward Fault and are now in northern Sonoma County. So in a sense, the topography has reversed: where East Bay Hills are today, there was a lowland; where the valley of San Francisco Bay is today was a highland.

Volcanoes!

About 10 million years ago a new geologic episode began. The southern East Bay was the site of volcanic activity related to the beginning of movement along the Hayward Fault in this area. By that time most of the East Bay was already land, but the present East Bay Hills had not yet started to rise. For almost a million years, lava and volcanic ash periodically covered the alluvial plains of the landscape. These volcanic rocks can be seen in the present-day hills above Berkeley and Oakland (pl. 118). They have been carried northward by continuing movement on the

Plate 118. Moraga Formation at Sibley Volcanic Regional Preserve, Oakland.

Hayward Fault. Basaltic flows called the Moraga Formation erupted from 10 to about 9 million years ago from Round Top in Robert Sibley Volcanic Regional Preserve and other vents. These lavas flowed over the alluvial plain in many separate flows. You can see them along Hwy. 24 east of the Caldecott Tunnel. There the lavas are steeply tilted, like the Orinda Formation, and form the prominent ridges on either side of the syncline (fig. 40). If traffic is slow, you will have time to see individual lava flows. They are several feet thick, and at the base of most is a red zone, called a baked contact, that formed as the hot lava flowed over damp alluvial sediments. The iron in the sediments and lava was oxidized by the hot lava to a bright red color. The lava flows are also exposed on Grizzly Peak Blvd. north of Fish Ranch Road in Berkeley. There you can stop and take a close look at excellent outcrops.

During the period of volcanic eruptions, one of the lava flows dammed a creek and formed a lake. Fine-grained light gray lake bed sediments, called the Siesta Formation, are present along Hwy. 24 at the Gateway Blvd. interchange between the Caldecott Tunnel and the Orinda exit. They are now folded into the center of the syncline. These lake bed sediments are soft rocks and easily eroded. Landslides have been common where they are exposed in the Gateway Blvd. area. About 9 million years ago another volcanic eruption took place nearby, forming the Bald Peak Basalt, which poured over the lake bed sediments.

As complicated as the East Bay geologic story is, the many excellent exposures of rocks throughout Alameda and Contra Costa County provide ample opportunities for you to unravel the past and to enjoy the varied landscape that has resulted from millions of years of geologic activity and change.

FROM A MAGNIFICENT FOGGY COASTLINE to the hot Central Valley, the three northern counties of the Bay Area—Sonoma, Napa, and Solano—are a microcosm of the Bay Area's complex geology (maps 26, 27). Explore the uncrowded back roads, stop at interesting road cuts, or walk the beaches along the coast to experience the wide range of processes that created the geologic mosaic of the North Bay. Along the coast, crashing waves and the San Andreas Fault greet you; volcanic hot springs beckon inland. And the dynamic early history of the Bay Area is revealed in rocks across the area.

Much of the North Bay is mountainous terrain with long ridges trending southeast to northwest, parallel to the San Andreas Fault. On the west the forest-covered mountains are interrupted only by the Russian River and by Dry Creek, its valley now partly filled by Lake Sonoma. A few narrow winding roads cross this rugged landscape, and respite from urban pressures comes quickly amid the splendid scenery and rocks. Along the coast, flat patches of marine terrace cling precariously to the edge of the mountains that plunge to the sea.

The Russian River cuts an unusual course through this landscape. From its headwaters in Mendocino County it runs southward about 55 miles toward Healdsburg, then makes a sharp bend to the west, flowing along the southern edge of the mountains to the sea at Jenner. The Russian River may once have flowed southward to the bay along the course followed today by the Petaluma River. A stream draining to the ocean in the vicinity of Healdsburg may have captured the Russian River and diverted its course to the west. Or it may have flowed westward before uplift of the mountains began and eroded down through the rocks fast enough to keep pace with the rising hills.

The mountains in the central and eastern parts of the North Bay are almost as rugged as those on the west. At 4,344 feet, Mount St. Helena, where the three North Bay counties meet, is the highest peak and the second highest in the Bay Area, after Mount Copernicus near Mount Hamilton in the South Bay. The long ridge of the Vaca Mountains rises to more than 3,000 feet at Berryessa Peak, which overlooks Lake Berryessa, a reservoir that is one of the chief recreational areas in the North Bay. The waters of Lake Berryessa flow into the Sacramento River through Putah Creek. The oak-woodland and grass-covered Vaca Mountains are largely beyond the reach of the coastal fogs and are dry and hot in comparison to those on the west.

Across the southern part of the North Bay, the landscape is one of lower, gentler hills separated by wide valleys, like long fingers reaching northward from San Pablo Bay. Along San Pablo and Suisun Bays, wetlands drained for agriculture in the nineteenth and twentieth centuries are being reclaimed and restored, home once again to a multitude of birds along the Pacific Flyway. They are part of the San Francisco Bay ecosystem and are discussed in chapter 6. The flat agricultural lands of the Central Valley make up the landscape east of the Vaca Mountains, along the I-80 corridor from Vacaville to the Yolo county line at Davis.

The varied and dramatic topography of the North Bay is young, and it is shaped by the same geologic processes as in much of the rest of the Bay Area. Tectonic movements fold and fault the landscape, erosion wears the high places down, and faults rearrange the rocks.

The Sonoma Coast: Lifted Up and Slumping Down

The Russian River divides the Sonoma coast into two dramatically different landscapes. To the south, it is gentle and rolling like that of Marin County. This landscape is underlain by soft, easily eroded rocks, in part Franciscan mélange (described in chapter 3), and in part marine sediments deposited in a shallow sea that covered southwestern Sonoma County about 6 million years ago (see below). Erosion of these rocks has created a break in the Coast Ranges, a lowland that allows wind and fog to cool southern Sonoma County in summer.

Most of the southern Sonoma coastline is parkland, and there are many places to access the beaches and rocky headlands. Along the shore, low tide exposes tide pools on the wave-cut platform, the gently sloping surface carved into the rock as waves erode the sea cliffs (see chapter 1). Picturesque sea stacks, remnants of a former shoreline, are common where the rocks are resistant to
text continues on page 260

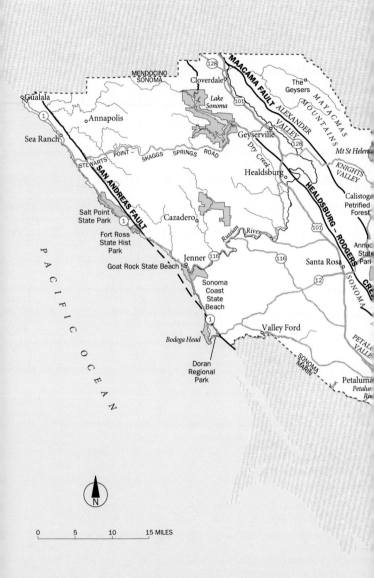

MENDOCINO
SONOMA

Gualala

Sea Ranch

Annapolis

Cloverdale

Lake
Sonoma

*The
Geysers*

MAACAMA FAULT

ALEXANDER
VALLEY

M A Y A C M A S

M O U N T A I N S

POINT – SKAGGS SPRINGS ROAD

128

101

Geyserville

Dry Creek

128

SAN ANDREAS FAULT

STEWARTS

Mt St Helena

KNIGHTS
VALLEY

Healdsburg

HEALDSBURG – RODGERS

Salt Point
State Park

Calistoga
Petrified
Forest

Cazadero

Fort Ross
State Hist
Park

1

Russian *River*

101

Annad
State
Park

Jenner

Goat Rock State Beach

116

116

Santa Rosa

CRE

Sonoma
Coast
State
Beach

12

SONOMA

PACIFIC

1

Bodega Head

Valley Ford

PETAL
VALLE

Doran
Regional
Park

SONOMA
MARIN

Petaluma
*Petalu
Riv*

OCEAN

N

0   5   10   15 MILES

Map 26. Map of the North Bay.

MENDOCINO
SONOMA

*Lake
Sonoma*

SAN ANDREAS FAULT

HEALDSBURG – RODGERS

MAACAMA FAULT

River

Russian

SONOMA
MARIN

P A C I F I C   O C E A N

N

0   5   10   15 MILES

	Quaternary alluvium, flood basin deposits in Delta
	Quaternary landslide deposits
	Quaternary beach and dune sand
	Quaternary Bay, Delta mud and marsh
	Quaternary sedimentary rocks
	Quaternary marine terrace deposits
	Quaternary/Tertiary sedimentary rocks
	Upper Tertiary sedimentary rocks
	Lower Tertiary sedimentary rocks
	Tertiary volcanic rocks
	Franciscan Complex coherent rocks
	Franciscan Complex mélange
	Great Valley Sequence
	Serpentinite, other Coast Range Ophiolite
	Salinian Complex granitic rocks

Map 27. Geologic map of the North Bay.

erosion (pl. 5). Highway 1 meanders across flat, uplifted marine terraces at the edge of the sea. These terraces are testimony to the dynamic geologic processes operating along this stretch of coast. They were formed at sea level 80,000 or 120,000 years ago as wave-cut platforms at times of high sea level during interglacial periods. In the short geological time since then, this coast has been uplifted to its present elevation by movement on the nearby San Andreas Fault. Notice that many of the larger sea stacks offshore have flat tops; they are remnants of the marine terrace.

When a wave-cut platform is uplifted to become a marine terrace, its sea stacks go up with it. At a number of places along the coast you can see former sea stacks jutting up from the present terrace and from older terraces to the east. At Goat Rock State Beach you can see several levels of marine terrace with their fossil sea stacks (see below).

The cliffs along the Sonoma coast are actively eroding back. Fortunately, most of the shore is undeveloped, and the battle between sea and land rages relatively unhindered. At Carmet-by-the-Sea north of Bodega Bay, however, the story is different. Here, as in Daly City (see chapter 7), homes that were built decades ago at the edge of the sea to take advantage of a splendid view have gradually lost their backyards to wave attack. Concrete sea walls constructed in an effort to slow erosion have been of little help. The winter storms of 1996/7 caused severe erosion, and in the storms of 2000, several homes slid to the beach below (pl. 119).

North of the Russian River from Jenner to Fort Ross, the coast is steep and wild, and the mountains drop straight to the sea. Hwy. 1 clings to the edge of the land; travelers on the ocean side of the road can look straight down, if they dare, to rocky headlands and pocket beaches hundreds of feet below. In strong winter storms when waves pound against the cliffs, you can feel the ground shake half a mile inland. This stretch of the coast is underlain by easily eroded Franciscan mélange, and the hillsides slump at an alarming rate in wet winters. Notice the hummocky look of the slopes, the fresh landslides, the new roadways and bridges, and the cracked old highway. Geologic processes are actively at work here. When the angle of the sun is low, you can see the wrinkles and scars of a constantly sliding landscape in what geologists call "melted ice cream topography" (pl. 120). Keeping this section of Hwy. 1 open through stormy winters is a major

Plate 119. Houses at edge of cliff at Carmet-by-the-Sea, Sonoma County.

Plate 120. "Melted ice cream topography" in Franciscan mélange along coast north of Jenner.

undertaking. Continual repairs are required, and increasingly sophisticated (and expensive) engineering structures are built to maintain the road in an area where probably no road should ever have been attempted.

From Fort Ross north to the Mendocino county line at Gua-

Plate 121. Sag pond filled with tules at Fort Ross.

Plate 122. Fort Ross State Historical Park. The fort sits on a marine terrace. The San Andreas Fault runs behind the grassy ridge at arrow in the middle of the photo.

lala, the mountains recede inland and remnants of a marine terrace characterize the coastal landscape. In this area the San Andreas Fault lies onshore. Fort Ross is a good place to consider the effects of the 1906 earthquake, including fences, roads, and an old orchard that have been offset by as much as 12 feet. Take a look at photographs in the visitor's center of Fort Ross State Historical Park and wander along the fault valley east of Hwy. 1. Fort Ross Creek has been offset by fault movement, and a little sag pond lies in the fault zone just north of Fort Ross Road (pl. 121). The reconstructed Russian fort at Fort Ross, which sits on a remnant of marine terrace (pl. 122), provides an interesting glimpse into the life of the fur-trading Russians on the California coast. The fort was founded in 1812 and was occupied until 1841. Its chapel, restored after the 1906 earthquake, is the oldest Russian Orthodox chapel still standing in the United States.

Faults and Fault Slices

Three major active fault zones transect the North Bay: the San Andreas along the coast, the Rodgers Creek in the central area, and the Green Valley in the east (map 27). All three are part of the San Andreas Fault System, and each of the faults has many branches. The Rodgers Creek Fault Zone includes the Healdsburg and Maacama Faults. The Rodgers Creek and Green Valley Faults may be part of the East Bay Fault System (see chapter 9); however, they do not connect directly with faults to the south, and the relations are still being debated.

The San Andreas and Rodgers Creek Faults and their branches divide the North Bay into many packets of rock, each with a somewhat different sequence of rocks and geologic history. Each is separated from its neighbors by presently active or ancient faults, and movement along these faults has brought the blocks together to their present configuration over time (see chapter 2). Because the dominant movement is right-lateral strike-slip, like that on the entire San Andreas Fault System, each packet moves northwestward relative to the one east of it (fig. 16). Where the faults bend or are offset, the crust is uplifted or downdropped, creating hills or valleys such as the Petaluma–Santa Rosa lowland.

The North Bay, like the rest of the Bay Area, is being squeezed by the transpression (strike-slip plus compressional movement) between the Pacific and North American Plates. This squeeze is shortening the crust and folding the area between the Rodgers

Creek and Green Valley Faults. Transpression, which has been going on for only a few million years, is one of the dominant processes shaping the young landscape of the North Bay.

Earthquake activity in the North Bay is concentrated along the Rodgers Creek and Green Valley Faults and their branches. The San Andreas Fault has been quiet since 1906, but there have been three significant earthquakes on the Rodgers Creek Fault in the past 4,000 years. The most recent activity was two moderate earthquakes of magnitude 5.6 and 5.7 in 1969 on it or a related fault. These earthquakes caused considerable damage in Santa Rosa, where the shaking was intense because of the deep alluvial fill in the valley. This fault is considered by seismologists to be one of the more likely northern California sections of the San Andreas Fault System to rupture. No major earthquakes have occurred on the Green Valley Fault since records have been kept, but it is actively creeping in the Cordelia area.

The Hole in the Head

Half a century ago the San Andreas Fault at Bodega Head was at the center of one of the first environmental battles in the United States. In the 1950s Pacific Gas & Electric Company (PG&E) proposed building a nuclear power plant on the granitic rocks of Bodega Head, a few hundred feet from the fault (pl. 123). At that time the concept of plate tectonics was unknown and the San Andreas Fault was not well understood. Coastal residents were first concerned about the impact of the plant on the area's fine scenery and recreational values. Then questions arose about the possibility of radioactive releases drifting across the west Marin and Sonoma dairy lands. Gradually, concern grew over the impact of an earthquake on the fault.

PG&E believed that a plant could be designed to withstand an earthquake and dug a large pit to house the reactor (the "hole in the head," as it became known). But before construction on the plant itself began, a visiting geophysicist toured the excavation and discovered a recently active strand of the San Andreas Fault crossing the site. This led to government concern about the safety of the plant. In 1964 PG&E withdrew its application for the plant and deeded the site to the county. Visiting it today, you would never guess its history. The former pit has become a freshwater

Plate 123. Site of proposed nuclear power plant at Bodega Head, looking east to town of Bodega Bay in 1976. The pit in foreground was dug to hold the reactor. Bodega Harbor in background lies in the San Andreas Fault Zone.

pond, and the construction site is now overgrown with coyote brush and other native vegetation. Stroll out on the boardwalk and muse about the origin of this pond at the edge of the San Andreas Fault.

Assembling the North Bay

The diverse rocks of the North Bay (map 27) record the geologic history of the area over the past 200 million years and more. The oldest rocks were formed when western California was still beneath the ocean. They record the plate collision, subduction, and accretion that took place during the Mesozoic and into the early Tertiary in this area. The youngest rocks are volcanic, erupted from volcanoes 10,000 years ago. Unconsolidated sediments form extensive marshes and mudflats along the southern margin of the North Bay and fill the area's stream valleys. They are the sedimentary rocks of the future.

The Basement Rocks

The oldest rocks in the North Bay, the basement rocks, are the same three types as in other parts of the Bay Area—the Salinian, Franciscan, and Great Valley Complexes (see chapter 2 for the plate collision story and chapter 3 for a description of the rocks). They are the rocks that were formed or assembled during the long episode of Mesozoic plate collision and subduction.

The Salinian Block

In the North Bay the Salinian Block consists of two parts, a southern section at Bodega Head and a long sliver to the north between Fort Ross and Point Arena in Mendocino County. The rocks on these two pieces are very different (map 27). At Bodega Head the basement rocks are Mesozoic granitic and older metamorphic rocks of the Salinian Complex with Quaternary dune sand and terrace deposits resting on them. Bodega Head, the northernmost site where Salinian Complex rocks can be found, is the above-water part of a ridge that extends north from Point Reyes. The granitic rocks were formed about 80 to 90 million years ago during Mesozoic subduction. They are similar to those at Tomales Point some 6 miles to the south, at the Farallon Islands, and at Montara Mountain on the Peninsula. Recent studies have shown that they also match rocks at Point Lobos on the Monterey Peninsula, separated by movement along the San Andreas Fault System.

The basement rocks between Fort Ross and Point Arena form the Point Arena Terrane of the Great Valley Complex. The rocks are mainly ocean crust basalt and marine sedimentary rocks deposited on the crust. The latter consist of many separate turbidity flows (see chapter 3) formed by sediment eroded from a continent, deposited in offshore marine basins, and then hardened into layers of conglomerate, sandstone, and shale. The layers, which were originally horizontal, have been steeply tilted in many places, such as at Salt Point State Park, which sits on a marine terrace north of Fort Ross. The rocks are wonderfully exposed in the sea cliffs in the park. Conglomerates and coarse sandstone of early Tertiary age can be seen near Gerstle Cove (pl. 124); finer-grained sandstone and shale occur to the north at Stump Beach. Along the trail between the two sites sits a fossil sea stack composed of conglomerate and sandstone. It was quarried

Plate 124. Tilted sandstone and shale of turbidity flow deposits beneath a marine terrace at Salt Point State Park.

for building blocks, and you can still see drill holes in the quarry face and in the blocks lying on the ground (pl. 125). Some of the shale layers contain the burrows of mud-dwelling organisms, such as worms and mollusks that foraged through the mud in search of food or shelter. Similar rocks can be seen at low tide on the beach at Schooner Gulch State Beach near Point Arena in Mendocino County (pl. 126).

Similar, but younger, turbidites are exposed in the sea cliffs along the shore at the Sea Ranch development, north of Salt Point. Sea Ranch sits on a wide marine terrace that extends for about 10 miles along the coast. Although the land is private property, you can get to the shore at several public access points. The volcanic ocean crust basalt on which the turbidites were deposited can be seen in the low sea cliffs at the Black Point access.

Across the Fault

East of the San Andreas Fault, the oldest rocks are the Franciscan and Great Valley Complex rocks that were formed and assembled during more than 100 million years of plate collision and subduction along the west coast of North America (see chapter 2).

Plate 125. Fossil sea stack of conglomerate on uplifted marine terrace, in Gerstle Cove, Salt Point State Park. Rock was quarried for building stone. Note drill holes in block in foreground.

Plate 126. Burrow at Schooner Gulch State Beach in Mendocino County.

The mountainous terrain between the San Andreas and Rodgers Creek Faults is dominantly Franciscan with some Great Valley Complex rocks faulted in (map 27). Between the Rodgers Creek and Green Valley Fault Zones, large masses of both types of basement occur in complexly faulted juxtaposition. East of the Green Valley Fault the basement rocks are largely Great Valley Complex with a few fault slivers of Franciscan. These rocks are described in chapter 3 and summarized below.

The Franciscan Complex

The Mesozoic Franciscan rocks (described in chapter 3) are similar to those found elsewhere in the Bay Area; however, subduction continued into the Tertiary in the North Bay (and in the Coast Ranges to the north), and turbidity currents continued to carry sediments into the subduction trench. These sediments formed the youngest Franciscan graywacke, which is present all along the Sonoma coast. The other typical Franciscan rock types are also found in the North Bay—pillow basalt, radiolarian chert, shale, and metamorphic rocks, as well as large areas of mélange. One of the few roads that traverse the mountainous landscape east of the San Andreas Fault, the Stewarts Point–Skaggs Springs Road between the coast and Healdsburg is a good place to see many of the Franciscan rocks. It takes you on a spectacular drive across a typical Franciscan Coast Ranges landscape.

The Franciscan rocks in the Coast Ranges north of San Francisco have been divided into three belts—eastern, central, and coastal—which differ in age and degree of metamorphism. All three are represented in the North Bay (map 28). The eastern belt includes the oldest and most altered Franciscan rocks; the coastal belt, the youngest and least altered. These belts have been further subdivided into terranes. Six Franciscan terranes and mélange have been mapped in the western half of the North Bay (map 28). Each terrane is a fault-bounded rock package that differs in either geologic history, type of rock, or sequence of rocks (table 2). The terranes occur as elongate packets that parallel the faults. Three of the terranes, the Yolla Bolly, Marin Headlands, and Novato Quarry Terranes, are present also in other parts of the Bay Area. Three are found only in the North Bay: the Cazadero, Devils Den Canyon, and Lake Sonoma Terranes.

The Cazadero Terrane includes excellent examples of the high-grade metamorphic rock blueschist, an altered basalt. At some sites you can see fine examples of the minerals that make up this rock, including silvery mica, bluish glaucophane, and long dark green crystals of actinolite. The Cazadero Terrane rocks are also interesting because they have been identified as a possible

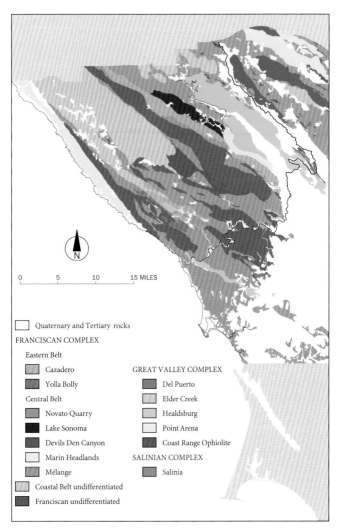

Map 28. Map of the Franciscan and Great Valley Complex Terranes in the western part of the North Bay.

Quaternary and Tertiary rocks

FRANCISCAN COMPLEX

Eastern Belt

Cazadero

Yolla Bolly

Central Belt

Novato Quarry

Lake Sonoma

Devils Den Canyon

Marin Headlands

Mélange

Coastal Belt undifferentiated

Franciscan undifferentiated

GREAT VALLEY COMPLEX

Del Puerto

Elder Creek

Healdsburg

Point Arena

Coast Range Ophiolite

SALINIAN COMPLEX

Salinia

source of pebbles in the Orinda Formation of the East Bay Hills (see chapter 9). When the sediments of the Orinda Formation were deposited about 12 million years ago, the Cazadero rocks formed a highland where San Francisco Bay is today. Since that time, movement along the Hayward, Rodgers Creek, and related faults has carried them about 60 miles to the northwest. A good place to see the blueschists is along the road between Cazadero and Fort Ross north of the Russian River.

The Marin Headlands Terrane, with pillow basalt, chert, and graywacke, occurs in the Geysers area, as well as along Dry Creek Valley southeast of Lake Sonoma. Although these rocks are far from the exposures in the Marin Headlands, the tiny radiolarians that make up much of the rock are the same type and age as those to the south. They are part of the widespread distribution of this terrane. Small outcrops of the Yolla Bolly Terrane are present near Berryessa Lake.

South of the Russian River the best place to see Franciscan rocks is along the beaches and sea cliffs of the Sonoma coast because inland they are covered by younger sedimentary rocks. For example, Shell Beach has beautiful wave-polished pebbles in many colors. Hard and resistant cherts make especially beautiful pebbles in shades of red, brown, yellow, and green. When the red and yellow colors are sharply defined, the rock is called jasper, a semiprecious stone. Along the coast you can also see Franciscan mélange with a variety of resistant Franciscan rock types embedded in the ground-up mélange matrix.

The Great Valley Complex

Both the ocean crust and mantle rocks of the Coast Range Ophiolite and the overlying marine sediments of the Great Valley Sequence (see chapter 3 for description) are represented in the North Bay (map 27). These rocks were formed during the 100 million years of plate collision and subduction in the Mesozoic. In Napa and Solano Counties they are part of the long linear belt of Great Valley Complex rocks that extends northward along the east side of the Coast Ranges from Lake Berryessa almost to the northern end of the Central Valley.

The ophiolite includes both flow and pillow basalts and serpentinite, which is considerably more common in the northeastern part of the North Bay than elsewhere in the Bay Area. In some road cuts, such as along the Berryessa-Knoxville Road north of

Plate 127. Great Valley Sequence turbidites at Monticello Dam at Lake Berryessa tilted almost vertical. Resistant sandstone forms ridge on left; grassy shale slopes are on far right.

Lake Berryessa, the serpentinite is shiny and polished into slickensides. In other places, you can see serpentinite that is highly fractured and piled up in a heap of rubble. This serpentinite is not part of the ophiolite but is derived from it, a sedimentary serpentinite consisting of eroded and redeposited rock. Both types of serpentinite make a thin, rocky soil that often is a deep red, in fine contrast to the bluish green of the rock. Another large area of serpentinite is present south of the Stewarts Point–Skaggs Springs Road in an area called "The Cedars." The trees on this serpentinite soil are not cedar, but cypress *(Cupressus sargentii)*.

The marine sedimentary rocks of the Great Valley Sequence in the North Bay consist of many individual thick and thin layers of sandstone and shale that were formed by turbidity flows (pl. 127). Uplift of the Vaca Mountains in Solano County has pushed up and tilted these layers into long ridges and valleys like those along the west side of the Sacramento Valley to the north (pl. 25). More resistant sandstones underlie the ridges and softer shales erode to form the valleys. Because of the way they have been tilted up, the youngest rocks are on the east, the oldest on the west. If you were to hike across these ridges and valleys westward from the Sacramento Valley to Lake Berryessa, you would walk back through time into the late Jurassic, some 150 million years ago.

The best exposures of turbidites are just east of the Bay Area in the valley of Cache Creek along Hwy. 16 between Rumsey and Hwy. 20 in the northwest corner of Yolo County. Here, Cache Creek and its tributary, Bear Creek, have sliced through layer after layer of turbidites. For miles you can see resistant light sandstone layers alternating with layers of dark shale, all tilted to a steep angle. The sandstone layers stick out like ribs from the sides of the valley.

Four Great Valley Sequence Terranes and the Coast Range Ophiolite have been mapped in the western half of the North Bay (map 28; see table 2 for terrane descriptions). A few miles east of Jenner, the Russian River has cut through a belt of Healdsburg Terrane conglomerate like the Novato Conglomerate found at Black Point in Marin County. Another large outcrop of this terrane forms the hills west of the Alexander Valley. The extensive outcrops of the Great Valley Sequence in the Vaca Mountains are Elder Creek Terrane. A few scattered outcrops of Del Puerto Terrane are also present in the North Bay.

After Subduction Ended

The change from a colliding to a sliding plate boundary marks the beginning of a new phase in the North Bay geologic story. Plate collision continued in the North Bay as the San Andreas Fault System propagated northward (see chapter 2). In the North Bay Franciscan graywackes were deposited in the subduction trench well into the Tertiary, until about 35 to 40 million years ago. Then, local basins replaced the trench and a different type of marine sediment accumulated.

In the southeastern part of the North Bay in Solano County, the older Tertiary rocks are a continuation of the same rock units found at Mount Diablo and Black Diamond Mines (see chapter 9). They record a time when alternately deep and shallow marine basins occupied this area, as a result of tectonic movements and sea level change. The Domengine, Nortonville, and Markley Formations make up the low hills north of Grizzly Bay. Caltrans has

created fine road cuts in the Markley sandstone at the east end of
Jamieson Canyon along Hwy. 12 west of Cordelia. Along Pleas-
ants Valley Road west of Vacaville, you can get a sense of the difference that rock hardness makes to the landscape. The road goes
through a valley between erosion-resistant Great Valley Sequence
rocks that form the high ridge of the Vaca Mountains on the west
and the much weaker Tertiary sedimentary rocks in the low hills
on the east. This is a delightful drive, especially in spring when
the apricot trees are in bloom.

Plate 128. Wilson Grove Formation sandstones near Valley Ford,
Sonoma County.

By about 8 to 9 million years ago, much of the eastern North
Bay had emerged from the sea, and the shoreline was approximately in central Sonoma County. To the east lay a broad coastal
plain and gentle highlands. Streams drained the highlands and
flowed westward across the plain to the sea, depositing sediments
in environments ranging from terrestrial on the east to shallow
marine on the west. Over time the terrestrial, estuarine, and delta
deposits hardened into rocks called the Petaluma Formation,
which today lies under both the surface sediments of the Santa
Rosa Valley and the Sonoma Volcanics to the east. The sediment
deposited in the shallow sea became marine sandstone that eventually hardened into the rocks called the Wilson Grove Forma-

tion, which underlies the hilly country west of Santa Rosa. They form a well-drained soil ideal for apple orchards and berries. These rocks are poorly consolidated and are easily eroded. You can see them only at occasional road cuts, such as along Hwy. 1 near Valley Ford (pl. 128). Along its eastern margin, the Wilson Grove marine sandstone interfingers with terrestrial rocks of the Petaluma Formation, an alternation of marine and terrestrial deposits that records changing sea level or tectonic movement of the land.

Between 3 and 4 million years ago a marine embayment covered part of northwestern Sonoma County (east of the San Andreas Fault) from the vicinity of Fort Ross north to the Mendocino county line. The Salinian Block had not yet traveled this far north, and open ocean lay to the west of the fault. The marine sediments deposited in this shallow sea formed the sandstone and conglomerate of the Ohlsen Ranch Formation. They are now more than 1,000 feet above sea level, uplifted by folding and faulting of the Coast Ranges in the past million years or so.

The North Bay's Volcanoes

If you are interested in volcanic rocks, the North Bay is an excellent place to see a large variety of volcanic landscapes and rocks (map 27). About 8 to 9 million years ago volcanoes started to erupt in the North Bay and over time covered much of the landscape with thick layers of lava and ash. The Sonoma Volcanics, as they are known (pl. 28), erupted intermittently in the central part of the North Bay until about 2.5 million years ago. After a period of quiet, volcanoes began to erupt again about 2 million years ago farther north in the vicinity of Clear Lake and continued to erupt until about 10,000 years ago. They are the most recent eruptions in the Bay Area. Most of these volcanoes erupted along faults related to the development of the San Andreas Fault System as it propagated northward. Fractures in the crust acted as conduits along which magma rose to the surface. The Sonoma and Clear Lake Volcanics continue the general trend of a northward progression of volcanic centers related to the transition from subduction to a sliding plate boundary in northern California (see chapter 2).

Plate 129. The Palisades east of the Napa Valley.

The Sonoma Volcanics consist of several types of volcanic rocks, including lava, mud flows, and tuff (see chapter 3). Because many of these rocks are resistant to erosion, they are a prominent part of the North Bay landscape. The highest peak in the North Bay, Mount St. Helena (4,343 feet; see chapter opening photo), is also volcanic rock but is not a volcano, although it looks rather like one. Most of the other high peaks in the area, such as Veeder Mountain (2,677 feet) and Atlas Peak (2,663 feet), are also composed of volcanic rocks. They form the mountain ridges framing the Petaluma–Santa Rosa, Sonoma, and Napa Valleys and the flanks of the Mayacmas Mountains in northeastern Sonoma County. The rocks on the ridge north of the Napa Valley, the Palisades, are volcanic (pl. 129); they and the volcanic rocks of Mount St. Helena are about 3 million years old.

Many interesting volcanic features can be explored in the Sonoma Volcanics. At Petrified Forest near Calistoga (pl. 130), an eruption took place about 3.2 million years ago like the one that we witnessed in 1980 at Mount St. Helens, Washington, when horizontal blasts blew down a forest. At Calistoga, the blast came from the direction of Mount St. Helena and leveled a forest of redwoods and firs. The redwoods were of a species *(Sequoia langsdorfi)* similar to the redwoods in the Bay Area today. As groundwater dissolved minerals out of the volcanic deposits, the living tissue of the trees was replaced by silica minerals, and the trees turned to stone. You can walk around these petrified trees in a private park west of Calistoga.

Plate 130.
Fallen tree at the
Petrified Forest.

Obsidian, the volcanic glass used by Native Americans to make arrowheads and spear points (pl. 131, *top*), is found at several localities in the Sonoma Volcanics. Glass Mountain, in the northern Napa Valley, has a high-quality obsidian and was a major source of material for tools. This obsidian, which is about 2.5 million years old, is exposed in a steep road cut on the Silverado Trail on the east side of the Napa Valley, half a mile north of Deer Park Road in St. Helena (pl. 131, *bottom*). Pieces of the rock look whitish at first, the result of weathering, but break one open and inside is beautiful black obsidian. Both the Glass Mountain obsidian and a deposit in what is now Annadel State Park east of Santa Rosa were quarried by Native Americans. More recently, in the nineteenth and twentieth centuries, basaltic and andesitic lavas in Annadel State Park were quarried for paving blocks and building materials. North of Glass Mountain on the Silverado Trail you can see a fine example of a baked contact (pl. 132), where an ash flow poured over damp ground and oxidized the iron minerals in the underlying material.

Plate 131. *Top*, Obsidian. *Bottom*, Glass Mountain road cut in obsidian of the Sonoma Volcanics.

Plate 132. Road cut in Sonoma Volcanics showing red baked contact at base, overlain by obsidian (dark layer on left) and weathered obsidian that has begun to crystallize. North of Deer Park Road on east side of Silverado Trail.

Lava, mud flows, and ash poured over the North Bay landscape for millions of years, but not continuously. Throughout much of the area, deposits of Sonoma Volcanics are interbedded with the Wilson Grove and Petaluma Formations described above. Fossils in these rocks tell us that in between volcanic outbursts large mammals roamed the highlands, plains, and shoreline of the North Bay, much as they did in the East Bay at this time.

Several volcanic units in the North Bay commonly have been grouped with the Sonoma Volcanics, but recent investigations show that they are older and today are not located where they erupted. For example, the Burdell Mountain volcanic rocks north of Novato in Marin County (see chapter 4) are similar to the Quien Sabe Volcanics of the South Bay and have been faulted northward more than 100 miles since they erupted about 10 to 12 million years ago. Several smaller volcanic units in Sonoma County are displaced parts of the Moraga Formation in the East Bay Hills, offset along the Hayward–Rodgers Creek or other fault systems as much as 25 miles.

The Geysers

The Clear Lake Volcanics started erupting from a center to the north of the Sonoma Volcanics about 2 million years ago. Most of these younger volcanic rocks were deposited in Lake County, but a few are found in northeastern Sonoma County near The Geysers. Although there have been no volcanic eruptions from the Clear Lake Volcanics in the past 10,000 years, surface evidence indicates that hot magma is still present beneath the area. For more than a century Bay Area residents have taken the baths at numerous hot springs in northern Sonoma, Napa, and neighboring counties, and the geothermal field at The Geysers has produced electricity since the 1950s. In spite of the name, The Geysers are not true geysers like Old Faithful in Yellowstone, which erupt hot water; they are fumaroles, which erupt steam. These fumaroles spouted out of holes drilled and cased in 1922 in an early attempt by the city of Cloverdale to develop them for energy (fig. 41). The Geysers Resort nearby was founded in an area of hot springs.

To form a geyser or fumarole one needs the equivalent of a stove, a pot, a lid, and a supply of water. In geologic terms this

Figure 41.
The Geysers
in the 1920s.

means a steady and shallow supply of heat, a reservoir of fractured rock to hold the water, and a cap of impermeable, relatively unfractured rock that holds the water and heat, allowing pressure to build up. At The Geysers, a magma chamber more than 2 miles below the surface is the stove that heats the fractured rock. Rainwater percolates into the hot rock and is heated to steam. The pot lid is serpentinite, which keeps the steam under pressure. When wells are drilled through the rock, the steam is tapped to drive turbines that generate electricity, providing enough energy to power a city the size of San Francisco (pl. 133).

The rising magma and associated hydrothermal (hot water) fluids of the Sonoma and Clear Lake Volcanics concentrated a number of valuable minerals that have been mined in the area, including silver, gold, and mercury. The minerals were dissolved and concentrated as the hot fluids rose along fractures in the crust and were redeposited in veins in the rock. The North Bay and Clear Lake areas were second only to the New Almaden

Plate 133.
A power plant
at The Geysers
in 2003.

Plate 134. Hydrothermally altered Franciscan rocks at The Geysers. Note steam rising from hot springs (bottom center).

Quicksilver District in the South Bay in the production of mercury in the Bay Area.

The hydrothermal fluids and gases in The Geysers area also have a strong effect on the landscape. They have altered much of the surrounding Franciscan rock (pl. 134), making it very susceptible to landsliding. Many of the hillsides look as if rock and soil have turned to thick cream, and virtually every square foot of hillside exhibits the hummocky appearance of earthflows. After heavy winter storms the Cloverdale-Geysers Road along Big Sulphur Creek is littered with rocks, mud, and tree limbs, as the hillsides liquefy and flow down into the valley. The creek itself contributes to the problem by undercutting the steep hillsides along its banks when it is swollen with runoff. The Geysers area provides many examples of geologic processes in action.

Rivers and Bay

A very different landscape and geology await the explorer of the southeastern corner of the North Bay. Here, where Solano, Contra Costa, and Sacramento Counties meet, the Sacramento River is joined by the San Joaquin River, and together they flow through Carquinez Strait into the bay. Extensive wetlands and gentle hills replace the more rugged ridges and valleys to the north.

Driving through the low rolling landscape of the Montezuma and nearby Potrero Hills provides a North Bay experience unlike any other (pl. 135). The Potrero Hills are formed of Tertiary rocks like those to the south at Black Diamond Mines (see chapter 9). The Montezuma Hills are composed of younger rock. They are alluvial fan deposits eroded out of an older highland in the past million years or so, then uplifted and gently dissected by runoff and streams. Notice that the hilltops are mostly the same elevation, about 200 to 300 feet. The fan deposits, called the Montezuma Formation, consist of poorly consolidated gravel, sand, and silt, which you can see in road cuts along the few roads that traverse this area.

A drive through the grass-covered Montezuma Hills turns your car into a time machine and transports you back to the rural

Plate 135. The Montezuma Hills, looking northeast across the Sacramento River from Sherman Island.

Bay Area of 100 years ago. Here, California's legendary green and golden hills are host to cows, wind turbines, and a few Victorian farmhouses. In early spring, when the grass is lush, hundreds of baby goats and sheep frolic on the hills.

The nearby wetlands of Suisun Marsh and the Grizzly Island Wildlife Area are just a few thousand years old. They formed as sea level rose and flooded this area after the last ice age (see chapter 6). They are remnants of the once-extensive marshes that surrounded San Francisco Bay until 150 years ago. The North Bay wetlands are a symbol of the Bay Area's young and ever-changing geologic landscape.

A Special Place to Explore: Goat Rock State Beach

At Goat Rock State Beach just south of the Russian River, many of the dynamic geologic processes that form the coastal landscape come together—from the ceaseless attack of ocean against land to the forces that brought up bits of the uppermost mantle from deep below the earth's surface more than 100 million years ago.

Plate 136. The uplifted marine terrace south of Goat Rock. Fossil sea stacks on the terrace; present-day sea stacks in water.

Highway 1 between Bodega Bay and Goat Rock lies on a wide, flat marine terrace, carved in places by stream channels (the dips in the road). After you turn into the entrance to Goat Rock State Beach, pull over at the second small turnout on the left for an excellent view of the terrace to the south. It has been uplifted more than 100 feet above the pounding waves that carved it at sea level 80,000 or more years ago. Note the large rock perched on the terrace, a fossil sea stack uplifted along with the terrace (pl. 136). Below it at the water's edge are sea stacks forming today. To the east across the highway, you can see a number of large fossil sea stacks on still higher and older terraces.

Plate 137. Goat Rock, a sea stack.

Flat-topped Goat Rock (pl. 137) was once a sea stack connected to land by a sand bar that has now been filled to make a parking lot. To the south is Arched Rock (pl. 138), where the vigorous attack of ocean waves has formed a natural bridge. Decades—or perhaps centuries—from now the bridge will collapse before the onslaught of the waves, creating two companion sea stacks.

The rocks at Goat Rock State Park tell of dynamic geologic activity in the past. The road to the south parking lot near Goat

Plate 138. Arched Rock.

Plate 139. Serpentinite at Goat Rock State Beach.

Rock is cut into blue green serpentinite, the California state rock (pl. 139). Glitters of light reflect off mirror-smooth patches called slickensides. They are evidence of the powerful forces that faulted or squeezed the rock to the surface from the earth's upper mantle. These polished surfaces were formed as rock ground on rock.

Soft rock was crushed in the process, and harder rock was polished. In the cliff here you can see the result: rounded, resistant blocks surrounded by crushed serpentinite.

Above the restroom at the east end of the parking lot, a small fault cuts the cliff and separates the serpentinite from Franciscan graywacke on the south. Squiggly white lines in the graywacke originated as relatively straight quartz veins filling fractures in the rock; they are testimony to the extensive folding the rock has undergone. Both Goat Rock and Arched Rock are composed of this graywacke.

Our look at the geology of the North Bay brings us full circle, back to Marin County where our exploration of Bay Area geology began. I hope that this journey through the Bay Area has increased your appreciation of its rich geologic mosaic and your understanding of the processes that created it. Its complexity is a delight and a challenge to us all, scientist, resident, and visitor. Geology is a dynamic science. New scientific discoveries will provide insight into the Bay Area's geologic history and the nuances of the landscape as we continue to explore the remote corners of this remarkable region.

GLOSSARY

Accretion In plate tectonics, the addition of island arc or continental material to a plate during convergent or strike-slip motion.

Accretionary wedge Material from a subducting plate that has been scraped off and added to the upper plate.

Alluvial fan A fan-shaped deposit of sediment formed at the base of hills as the velocity of a stream decreases.

Alluvial plain A level or gently sloping area of stream deposits.

Alluvium Sediment deposited by a stream.

Andesite A volcanic rock of intermediate silica composition.

Antecedent stream A stream that maintains and incises its course across a surface that has been uplifted.

Anticline A convex-upward fold of rock layers.

Aquifer A layer of sand, gravel, or porous rock that holds a significant amount of water.

Ash flow An avalanche of hot volcanic gases and fine particles, which cools to form a type of tuff.

Asthenosphere The hot, plastic layer of the upper mantle below the lithosphere.

Baked contact The reddish oxidized zone formed when lava flows over a wet surface.

Basalt A volcanic rock rich in iron and magnesium and with low silica content.

Basement rock The older rocks in an area that are overlain by younger units.

Berm The horizontal or gently sloping portion of a beach.

Block Used in two senses: (1) a large unit bounded by faults that behaves as a coherent unit; (2) small and large rocks set in mélange.

Blueschist A metamorphic rock containing high-pressure and low-temperature minerals formed in a subduction zone.

Chert A fine-grained sedimentary rock composed of silica.

Complex A group of different rocks that are closely associated in the field.

Compression Stresses directed against each other that tend to shorten the crust by folding or faulting.

Conglomerate A sedimentary rock composed mainly of pebble-size or larger, rounded rocks.

Convergent plate boundary The boundary between two plates that are moving toward each other.

Cusp Evenly spaced points of sand formed on a beach.

Dacite A volcanic rock whose silica content is intermediate between rhyolite and andesite.

Debris slide A type of landslide in which material moves rapidly as a coherent mass.

Deposition The settling of sediment, as in a lake or stream.

Diabase A rock of basaltic composition that commonly forms dikes and sills.

Diatom A microscopic single-celled plant with a silica skeleton.

Dike A tabular body of magma that cuts across the surrounding rock.

Diorite A plutonic rock of intermediate composition.

Divergent plate boundary The boundary between two tectonic plates that are moving away from each other.

Earthflow A type of landslide in which material moves as a viscous liquid.

Earthquake The sudden release of energy when a fault ruptures.

Eclogite A metamorphic rock formed in the earth's mantle, composed of the minerals garnet and pyroxene.

Epicenter The point on the earth's surface directly above the place where an earthquake originates.

Erosion The transport of rock by gravity, water, or other means.

Estuary A partially enclosed body of water where freshwater and saltwater mix.

Fault A fracture in rock along which movement has taken place.

Fault creep Slow movement along a fault that occurs without an earthquake.

Feldspar A group of silicate minerals that make up much of the earth's crust.

Foraminifer A single-celled animal whose skeleton is made of calcium carbonate.

Fore-arc basin The marine basin between a volcanic arc and a subduction trench.

Formation A geologic unit of rocks related by origin or type.

Fossil Remains or traces of a plant or animal preserved in rock.

Fumarole A volcanic vent that emits steam and other gases.

Geyser A volcanic vent that emits hot water.

Gneiss A metamorphic rock with bands of light and dark minerals.

Graded bedding A sedimentary layer with coarse particles at the base, grading into finer particles at the top.

Granite A plutonic rock.

Granitic Referring to silica-rich rocks such as granite and granodiorite.

Granodiorite A plutonic rock of intermediate chemical composition.

Graywacke A type of sandstone containing grains of various sizes and compositions.

Greenstone A basalt that has been altered and colored by the presence of green minerals such as chlorite.

Hot spot A localized center of volcanic activity usually not related to a spreading center.

Igneous Rocks formed by the cooling of molten rock (magma).

Lava Molten rock (magma) that has erupted to and cooled at the surface.

Limestone A sedimentary rock composed of calcium carbonate.

Lithify To turn sediments into solid rock through compression or cementation or both.

Lithosphere The earth's rigid outer shell, consisting of the crust and the uppermost part of the mantle.

Longshore drift Movement of sediment parallel to the shoreline by waves striking at an angle.

Magma Molten rock.

Magma chamber An area in the crust where magma pools.

Magnitude A measure of the energy released in an earthquake.

Mantle The layer of the earth between the crust and the core.

Marble A metamorphic rock composed of calcium carbonate; metamorphosed limestone.

Marine terrace An uplifted wave-cut platform.

Massive Referring to sedimentary rock that occurs as thick beds.

Mélange A mixture of rock materials, consisting of small to very large blocks in a ground-up matrix.

Metamorphic A type of rock altered by heat or pressure or both.

Normal fault A fault in which one block moves down relative to another block.

Obsidian Volcanic glass.

Ophiolite A group of rocks consisting of upper mantle, ocean crust, and ocean-floor sediments.

Outcrop Rock exposed at the surface.

Peat An organic deposit that consists of the partially decomposed remains of plants.

Pillow lava Rounded, bulbous lava flows formed as lava erupts underwater.

Plate tectonics The concept that the earth's lithosphere consists of rigid, mobile plates.

Plutonic A type of igneous rock formed as magma cools below the surface.

Quartz A mineral composed of silicon and oxygen.

Quartzite A rock composed of metamorphosed sandstone.

Radiolaria Microscopic single-celled animals with silica skeletons.

Radiolarian chert Chert composed largely of the skeletons of Radiolaria.

Reverse fault A fault in which one block moves at an angle over another block.

Rhyolite A volcanic rock with a high silica content.

Rockfall A type of landslide composed of rocks falling by gravity.

Sag pond A water-filled depression along a fault zone.

Sandstone A sedimentary rock composed of sand grains.

Schist A metamorphic rock with foliation and platy minerals such as mica.

Seamount A submarine mountain, commonly volcanic.

Sedimentary A type of rock formed from particles of other rock, organic particles, or precipitation from solution.

Seismology The study of earthquakes.

Serpentinite A metamorphic rock composed of the mineral serpentine; the California state rock.

Shale A fine-grained sedimentary rock composed of mud particles.

Silica Silicon dioxide, SiO_2; the chemical basis for many minerals that make up the earth's crust.

Slate A fine-grained metamorphic rock; metamorphosed shale.

Slickenside A polished surface produced by rock rubbing against rock under pressure.

Slump A type of landslide in which coherent material rotates as it moves down the slope.

Spreading center A plate boundary where two plates diverge or spread apart.

Strike-slip fault A fault with lateral movement, with one block sliding past another.

Subduction The sinking of one tectonic plate beneath another.

Syncline A down fold.

Tafoni Cavelike holes eroded into rock, also called honeycomb weathering.

Tectonic Referring to movements of the crust.

Terrane A fault-bounded body of rock, characterized by a geological history different from that of neighboring terranes.

Thrust fault A reverse fault in which one block moves over another block at a low angle.

Topography The physical relief of an area.

Transform plate boundary A plate boundary at which one plate slides past another.

Transpression Stress that includes both compression and sliding (transform) movement.

Transtension Stress that includes both extension and sliding (transform) movement.

Tuff A volcanic rock formed from fine volcanic particles.

Turbidite A sedimentary rock formed from the deposits of a turbidity current.

Turbidity current A mixture of sediment and water that flows down the continental slope to the deep ocean floor.

Two-layer circulation The flow of saltwater upstream beneath the freshwater layer in an estuary.

Unconsolidated Loose soil or sediment that is not compressed or cemented together.

Vein A deposit of minerals in a thin fracture in rock; commonly referring to a deposit of ore minerals.

Vesicle A small cavity in a volcanic rock formed by escaping gas.

Volcanic A type of igneous rock formed as magma cools at the earth's surface.

Volcanic Arc A chain of volcanoes, such as the Aleutian Islands, that forms as a subducting plate begins to melt.

Watershed The area drained by a stream.

Wave-cut platform A horizontal or gently sloping surface cut by waves near the shoreline.

Weathering The breakdown of rock by mechanical or chemical processes.

FURTHER READING

Published Works

Bailey, E. H., W. P. Irwin, and D. L. Jones. 1964. *Franciscan and related rocks, and their significance in the geology of western California.* Bulletin 183. San Francisco: California Division of Mines and Geology.

Bedrossian, T. 1974. Geology of the Marin Headlands, California Division of Mines and Geology. *California Geology* 27 (4):75–86.

Blake, M. C. Jr., ed. 1984. *Franciscan geology of northern California.* Book 43. Los Angeles: Pacific Section, Society of Economic Paleontologists and Mineralogists (now SEPM).

Bonilla, M. G. 1965. Geologic map of the San Francisco south quadrangle, California. U.S. Geological Survey open-file map, scale 1:20,000.

Brabb, E. E., F. A. Taylor, and G. P. Miller. 1982. Geologic, scenic, and historic points of interest in San Mateo County, California. U.S. Geological Survey map, I-1257-B, scale 1:62,500.

Chin, J. L., F. L. Wong, and P. R. Carlson. 2004. *Shifting shoals and shattered rocks—How man has transformed the floor of west-central San Francisco Bay.* Circular 1259. Reston, Va.: U.S. Geological Survey.

Clary, R. H. 1980. *The making of Golden Gate Park, San Francisco.* San Francisco: California Living Books.

Collier, M. 1999. *A land in motion: California's San Andreas Fault.* Berkeley and Los Angeles: University of California Press.

Harden, D. R. 2004. *California geology,* 2nd ed. Upper Saddle River, N.J.: Prentice Hall.

Hough, S. E. 2004. *Finding fault in California: an earthquake tourist's guide.* Missoula, Mont.: Mountain Press Publishing Company.

Howard, A.D. 1979. *Geologic history of middle California.* Berkeley and Los Angeles: University of California Press.

Jenkins, O.P., ed. 1951. *Geologic guidebook of the San Francisco Bay counties.* Bulletin 154. San Francisco: California Division of Mines.

Karl, H. A., J. L. Chin, E. Ueger, P. H. Stauffer, and J. W. Hendley II. 2001. *Beyond the Golden Gate: Oceanography, geology, biology, and environmental issues in the Gulf of the Farallones.* Circular 1198. Reston, Va.: U.S. Geological Survey.

Kious, W. J., and R. I. Tilling. 1996. *This dynamic planet: The story of plate tectonics.* Washington, D.C.: U.S. Geological Survey.

Konigsmark, T. 1998. *Geologic trips, San Francisco and the Bay Area.* Gualala, Calif.: GeoPress.

Lillie, R. J. 2005. *Parks and plates—The geology of our national parks, monuments, and seashores.* New York: W.W. Norton and Co.

McPhee, J. 1993. *Assembling California.* New York: Farrar, Straus, Giroux.

Moores, E. M., D. Sloan, and D. L. Stout, eds. 1999. *Classic Cordilleran concepts: A view from California.* Special Paper 338. Boulder, Colo.: Geological Society of America.

Norris, R.M., and R.W. Webb. 1990. *Geology of California,* 2nd ed. New York: John Wiley & Sons.

Oakeshott, G.B. 1978. *California's changing landscapes,* 2nd ed. New York: McGraw-Hill.

Parsons, Tom, ed. 2002. *Crustal structure of the coastal and marine San Francisco Bay Region, California.* U.S. Geological Survey Professional Paper 1658.

Schlocker, J. 1974. *Geology of the San Francisco north quadrangle, California.* U.S. Geological Survey Professional Paper 782.

Sloan, D., and D.L. Wagner, eds. 1991. *Geologic excursions in northern California: San Francisco to the Sierra Nevada.* Special publication 109. Sacramento: California Division of Mines and Geology.

Stoffer, P.W. 2005. *The San Andreas Fault in the San Francisco Bay Area, California.* U.S. Geological Survey Open-File Report 2005-1127. http://pubs.usgs.gov/of/2005/1127/ (accessed August 2005).

Stoffer, P.W., and L.C. Gordon, eds. 2001. *Geology and natural history of the San Francisco Bay Area: A field-trip guidebook.* U.S. Geological Survey Bulletin 2188. http://geopubs.wr.usgs.gov/bulletin/b2188/ (accessed August 2005).

Swinchatt, J. and D.G. Howell. 2004. *The winemaker's dance: explor-

ing terroir in the Napa Valley. Berkeley and Los Angeles: University of California Press.

Wagner, D.L., and S.A. Graham, eds. 1999. *Geologic field trips in Northern California.* Special publication 119. Sacramento: California Division of Mines and Geology.

Wahrhaftig, C. 1984. *A streetcar to subduction,* rev. ed. Washington, D.C.: American Geophysical Union.

Wallace, R.E., ed. 1990. *The San Andreas Fault System, California.* U.S. Geological Survey Professional Paper 1515.

U.S. Geological Survey Web Sites

Bay Area geologic maps, http://wrgis.wr.usgs.gov/wgmt/sfbay/geomap.html (accessed August 2005).

Earthquake information, http://quake.wr.usgs.gov/ (accessed August 2005).

Landslide information, http://sfslide.wr.usgs.gov/ (accessed August 2005).

San Francisco Bay and Delta information, http://sfbay.wr.usgs.gov (accessed August 2005).

San Francisco Bay Area regional database (BARD), http://bard.wr.usgs.gov/ (accessed August 2005).

San Francisco Bay Region geology, http://sfgeo.wr.usgs.gov/ (accessed August 2005).

Other Web Sites

American Geological Institute, www.agiweb.org (accessed August 2005).

Association of Bay Area Governments, Bay Area shaking hazard maps, www.abag.ca.gov/bayarea/eqmaps/mapsba.html (accessed August 2005).

Black Diamond Mines Regional Preserve, East Bay Regional Park District, http://ebparks.org/parks/black.htm (accessed August 2005).

California Geological Survey (formerly Division of Mines and Geology), www.consrv.ca.gov/cgs (accessed August 2005).

The California Geotour: An index to online geologic field trip guides of California, www.conservation.ca.gov/cgs/geotour/ (accessed February 2006).

California state beaches and parks, www.parks.ca.gov/ (accessed August 2005).

Geological Tour of UC Berkeley: Explore more!, www.seismo.berkeley.edu/seismo/geotour/ (accessed February 2006).

Information on Santa Clara County parks, www.parkhere.org (accessed August 2005).

John Karachewski's Geoscapes Photography, http://www.geoscapesphotography.com (accessed October 2005).

Mount Diablo geology, www.mdia.org/geology.htm (accessed August 2005).

Northern California Geological Society, www.ncgeolsoc.org (accessed August 2005).

Oakland Museum, Bay Area creeks information, http://museumca.org/creeks (accessed August 2005).

Peninsula regional parks: Mid-Peninsula Regional Open Space District, www.openspace.org (accessed August 2005).

San Francisco Estuary Institute, www.sfei.org (accessed August 2005).

San Francisco Estuary project information on water quality, wetlands, and resources of the bay and Delta ecosystem, www.abag.ca.gov/bayarea/sfep (accessed August 2005).

FIGURE AND MAP SOURCES

FIG. 1   After J. W. Downs, F. K. Lutgens, and E. J. Tarbuck, *Essentials of Geology,* 8th ed. (Upper Saddle River, N.J.: Prentice Hall, 2003).

FIG. 2   Courtesy of U.S. Geological Survey.

FIG. 3   After G. E. Weber and A. O. Allwardt, "The Geology from Santa Cruz to Point Año Nuevo: The San Gregorio Fault Zone and Pleistocene Marine Terraces," in P. W. Stoffer and L. C. Gordon, eds., *Geology and Natural History of the San Francisco Bay Area: A Field-Trip Guidebook,* U.S. Geological Survey Bulletin 2188, 2001, http://geopubs.wr.usgs.gov/bulletin/b2188/ (accessed August 2004).

FIG. 4   After E. J. Tarbuck and F. K. Lutgens, *Earth Science,* 10th ed. (Upper Saddle River, N.J.: Prentice Hall, 2003); D. McGeary, C. C. Plummer, and D. H. Carlson, *Earth Revealed,* 5th ed. (Boston: McGraw-Hill, 2003).

FIG. 5   After W. J. Kious and R. I. Tilling, *This Dynamic Earth: The Story of Plate Tectonics* (Washington, D.C.: U.S. Geological Survey, 1996); C. C. Plummer and D. McGeary, *Physical Geology,* 5th ed. (Dubuque, Iowa: Wm. C. Brown, 1991).

FIG. 6   After P. W. Stoffer and L. C. Gordon, eds., *Geology and Natural History of the San Francisco Bay Area: A Field-Trip Guidebook,* U.S. Geological Survey Bulletin 2188, 2001, http://geopubs.wr.usgs .gov/bulletin/b2188/ (accessed August 2004); Plummer and McGeary (see fig. 5).

FIG. 7   Courtesy of U.S. Geological Survey.

FIG. 8   After Plummer and McGeary (see fig. 5).

FIG. 9   After F. Press and R. Siever, *Understanding Earth,* 4th ed. (New York: W. H. Freeman, 1994).

FIG. 10 After C. W. Montgomery and D. Dathe, *Earth: Then and Now,* 2nd ed. (Dubuque, Iowa: Wm. C. Brown, 1994).

FIG. 11 Courtesy of Berkeley Seismological Laboratory.

FIG. 12 After R. E. Wallace, ed., *The San Andreas Fault System, California,* U.S. Geological Survey Professional Paper 1515, 1990.

FIG. 13 After Wallace (see fig. 12).

FIG. 15 After D. R. Harden, *California Geology* (Upper Saddle River, N.J.: Prentice Hall, 1997); E. A. Keller and N. Pinter, *Active Tectonics: Earthquakes, Uplift, and Landscape* (Upper Saddle River, N.J.: Prentice Hall, 1997).

FIG. 16 After Harden (see fig. 15).

FIG. 17 After Montgomery and Dathe (see fig. 10).

FIG. 18 Courtesy of Benita Murchey, U.S. Geological Survey.

FIG. 19 After W. K. Hamblin, *Earth's Dynamic Systems,* 5th ed. (New York: Macmillan, 1989).

FIG. 22 Courtesy of U.S. Geological Survey.

FIG. 24 Based on U.S. Coast and Geodetic Survey Chart 18649, San Francisco entrance.

FIG. 25 After A. Cohen, *An Introduction to the San Francisco Estuary,* 3rd ed. (San Francisco: Save The Bay, San Francisco Estuary Project, San Francisco Estuary Institute, 2000).

FIG. 26 After T. J. Conomos, ed., *San Francisco Bay: The Urbanized Estuary* (San Francisco: Pacific Division, American Association for the Advancement of Science, 1979).

FIG. 27 After Cohen (see fig. 25).

FIG. 28 After Cohen (see fig. 25).

FIG. 29 After figure by Kat G. Kalamars, based on sketch by Walt Hensolt.

FIG. 30 After B. Atwater, B. E. Ross, and J. F. Wehmiller, "Stratigraphy of Late Quaternary Estuarine Deposits and Amino Acid Stereochemistry of Oyster Shells beneath San Francisco Bay, California," *Quaternary Research* 16 (1981): 181–200.

FIG. 31 Courtesy of U.S. Geological Survey.

FIG. 32 Courtesy of U.S. Geological Survey.

FIG. 33 Courtesy of the San Francisco Estuary Institute.

FIG. 34 Courtesy of the Bancroft Library/University of California, Berkeley.

FIG. 35 After T. I. Iwamura, "Hydrogeology of the Santa Clara and Coyote Valleys Groundwater Basins, California," in E. M. Sanginés, D. W. Andersen, and A. V. Buising, eds., *Recent geologic studies in the San Francisco Bay Area,* vol. 76 (Fullerton, Calif.: Pacific Section, Society of Economic Paleontologists and Mineralogists [now SEPM], 1995).

FIG. 36 Courtesy of U.S. Geological Survey.

FIG. 37 Courtesy of New Almaden Quicksilver Mining Museum and New Almaden Quicksilver County Park Association, photo 1999-001-001.

FIG. 38 From A. C. Lawson and C. Palache, "The Berkeley Hills: A Detail of Coast Range Geology," *University of California Bulletin of the Department of Geology, Berkeley* 2 (1902): pl. 10.

FIG. 39 From R. A. Stirton, "Prehistoric Land Animals of the San Francisco Bay Region," in O. P. Jenkins, ed., *Geologic Guidebook of the San Francisco Bay Counties,"* Bulletin 154 (San Francisco: California Division of Mines, 1951).

FIG. 40 Courtesy of U.S. Geological Survey.

FIG. 41 Courtesy of the Bancroft Library/University of California, Berkeley.

MAP 3 Courtesy of California Geological Survey.

MAP 6 After C. M. Wentworth, *General Distribution of Geologic Materials in the San Francisco Bay Region, California: A Digital Database,* U.S. Geological Survey Open-File Report 97-744, scale 1:275,000, 1997, http://wrgis.wr.usgs.gov/open-file/of97-744/ (accessed January 2005).

MAP 8 J. C. Clark and E. E. Brabb, *Geology of Pt. Reyes National Seashore: A Digital Database,* U.S. Geological Survey Open-File Report 97-456, scale 1:48,000, 1997, http://wrgis.wr.usgs.gov/open-file/of 97-456/ (accessed January 2005).

MAP 9 J. C. Clark, C. Wahrhaftig, and E. E. Brabb, "San Francisco to Pt. Reyes: Both Sides of the San Andreas Fault," in D. Sloan and D. L. Wagner, eds., *Geologic Excursions in Northern California,* Special Publication 109 (Sacramento: California Division of Mines and Geology, 1991).

MAP 10 After W. P. Elder, "Geology of the Golden Gate Headlands," in P. W. Stoffer and L. C. Gordon, eds., *Geology and Natural History of the San Francisco Bay Area: A Field-Trip Guidebook,* U.S. Geological Survey Bulletin 2188, 2001, http://geopubs.wr.usgs.gov/bulletin/b2188/b2188 (accessed January 2005); geology by Clyde Wahrhaftig, 1974–1984.

MAP 12 J. Schlocker, *Geology of the San Francisco North Quadrangle, California,* U.S. Geological Survey Professional Paper 782, 1974; M. G. Bonilla, *Geologic Map of the San Francisco South Quadrangle, California,* U.S. Geological Survey Open-File Map, scale 1:20,000, 1965.

MAP 13 C. Wahrhaftig, *A Streetcar to Subduction,* rev. ed. (Washington, D.C.: American Geophysical Union, 1984).

MAP 21 D. L. Wagner, E. J. Bortugno, and R. D. McJunkin, *Geologic Map of the San Francisco–San Jose Quadrangle,* California Geological Survey Map 5A, scale 1:250,000, 1991.

MAP 25 After Mount Diablo Interpretive Association, *Trail Map of Mount Diablo State Park,* 4th ed, 1991.

PHOTO CREDITS

All photos by John Karachewski unless noted otherwise.

PLATE 1 Courtesy of U.S. Geological Survey.

PLATE 7 Photo by Margaret Gennaro.

PLATE 8 Photo by Robert Wiegel.

PLATE 14 Photos by Margaret Gennaro.

PLATE 16 Photo by Doris Sloan.

PLATE 18 *Top:* photo by Margaret Gennaro.

PLATE 19 Photo by Doris Sloan.

PLATE 21 Photo by Margaret Gennaro.

PLATE 25 Photo by Doris Sloan.

PLATE 46 Photo by Doris Sloan.

PLATE 48 Photo by Margaret Gennaro.

PLATE 54 Photo by Doris Sloan.

PLATE 58 Photo by John S. Shelton.

PLATE 59 Photo by Doris Sloan.

PLATE 67 Courtesy of People for Open Space; now Greenbelt Action.

PLATE 68 Photo by Doris Sloan.

PLATE 69 *Top:* photo by Doris Sloan.

PLATE 70 Photos by John S. Shelton.

PLATE 74 Photo by Margaret Gennaro.

PLATE 76 Photo by Clyde Wahrhaftig.

PLATE 78 Photo by Doris Sloan.

PLATE 79 Photos by Doris Sloan.

PLATE 90 Photo by Doris Sloan.

PLATE 100 Photo by Doris Sloan.

PLATE 101 Photo by Pacific Aerial Surveys, Oakland, California Photo K-ALA-C19K-43.

PLATE 102 Photo by Doris Sloan.

PLATE 115 Photo by Doris Sloan.

PLATE 119 Photo by Doris Sloan.

PLATE 123 Photo by Doris Sloan.

PLATE 125 Photo by Doris Sloan.

PLATE 126 Photo by Doris Sloan.

PLATE 128 Photo by Doris Sloan.

PLATE 131 *Top:* photo by Margaret Gennaro. *Bottom:* photo by Doris Sloan.

PLATE 137 Photo by Doris Sloan.

ADDITIONAL CAPTIONS

PAGE i Folded Franciscan chert, Marin Headlands.

PAGES ii–iii View of Mount Diablo and North Peak (right), looking west.

PAGE vi View north to Marin County from south tower of Golden Gate Bridge (photograph by Doris Sloan).

PAGES xviii–1 Satellite view of the Bay Area (photograph by NASA).

PAGES 24–25 Serpentinite, the California state rock, northeast of Bakers Beach, San Francisco; Golden Gate Bridge in background.

PAGES 46–47 Folded Franciscan blueschist at Sunol Regional Wilderness.

PAGES 76–77 Sunset over Bolinas Lagoon, looking toward Point Reyes.

PAGES 108–109 Sunrise view of San Francisco from Golden Gate Bridge.

PAGES 132–133 Aerial view over San Francisco Bay, toward northwest across Marin County; Angel Island in foreground, Tiburon Peninsula in center, Mount Tamalpais under clouds (photograph by Doris Sloan).

PAGES 156–157 Año Nuevo Beach on San Mateo coast.

PAGES 184–185 View to west from Mount Hamilton.

PAGES 216–217 View to northeast from Mount Diablo; in distance, Sacramento River (left) and San Joaquin River (right).

PAGES 252–253 Looking north toward Mount St. Helena across Knights Valley.

GEOLOGIC MAPS INDEX

FORMATIONS AND COMPLEXES	Bay Area, pp. 6–7	Marin County, pp. 82–83	Point Reyes, p. 89	Marin County Terranes, p. 96
Quaternary				
Olema Creek and Millerton Formation			●	
Tertiary				
Colma Formation				
Merced Formation			●	
Purisima Formation			●	
Santa Cruz Mudstone			●	
Santa Margarita Sandstone			●	
Monterey Formation			●	
Laird Sandstone			●	
Point Reyes Conglomerate			●	
Mesozoic Franciscan Complex				
Franciscan Complex coherent rocks	●	●		
Franciscan Complex mélange	●	●		
Franciscan pillow and flow basalt, greenstone				
Franciscan Complex basalt				
Franciscan Complex chert				
Franciscan Complex graywacke, shale				
Franciscan serpentinite and serpentite mélange				
Franciscan Complex limestone				
Franciscan Complex undivided			●	
Mesozoic Coast Range Complex				
Great Valley Sequence	●	●		
Coast Range Ophiolite diabase				
Coast Range Ophiolite pillow basalt				
Serpentinite, other Coast Range Ophiolite	●	●		●
Salinian Complex				
Granitic and metamorphic rocks	●	●		

Marin Headlands, p. 100	San Francisco, pp. 114-115	Franciscan Terranes, p. 124	The Peninsula, p. 161	South Bay, pp. 190-191	Diablo Range, p. 200	East Bay, pp. 222-223	Mount Diablo State Park, p. 240	North Bay, pp. 258-259	North Bay Terranes, p. 270
	•								
•	•		•	•		•		•	
•	•		•	•		•	•	•	
•	•								
							•		
•	•						•		
•	•								
	•								
				•					
					•				
			•	•	•	•	•	•	
							•		
							•		
			•	•	•	•	•	•	
			•					•	

Marin Headlands, p. 100	San Francisco, pp. 114–115	Franciscan Terranes, p. 124	The Peninsula, p. 161	South Bay, pp. 190–191	Diablo Range, p. 200	East Bay, pp. 222–223	Mount Diablo State Park, p. 240	North Bay, pp. 258–259	North Bay Terranes, p. 270
●	●		●			●			
●	●		●	●	●	●	●	●	
●	●			●		●	●	●	
			●	●					
						●		●	
●	●		●			●		●	
			●						
			●					●	
								●	
			●	●	●	●		●	●
									●
			●	●	●	●		●	
							●		
							●		
							●		
			●	●	●	●		●	
			●	●	●	●	●	●	

continued ➢

TERRANES	Bay Area, pp. 6–7	Marin County, pp. 82–83	Point Reyes, p. 89	Marin County Terranes, p. 96
Franciscan Complex				
Alcatraz terrane				
Cazadero terrane				
City College mélange*				
Coastal Belt undifferentiated				
Devils Den Canyon terrane				
Fort Point-Hunters Point serpentinite mélange*				
Franciscan undifferentiated				
Lake Sonoma terrane				
Marin Headlands terrane				●
Mélange*				●
Nicasio Reservoir terrane				●
Novato Quarry terrane				●
San Bruno Mountain terrane				●
Yolla Bolly terrane				●
Great Valley Complex				
Del Puerto terrane				
Elder Creek terrane				
Healdsburg terrane				●
Point Arena terrane				
Serpentinite and other Coast Range Ophiolite				●
Salinian Complex				
Salinia terrane				●

MAPS

* Informal Unit

Marin Headlands, p. 100	San Francisco, pp. 114–115	Franciscan Terranes, p. 124	The Peninsula, p. 161	South Bay, pp. 190–191	Diablo Range, p. 200	East Bay, pp. 222–223	Mount Diablo State Park, p. 240	North Bay, pp. 258–259	North Bay Terranes, p. 270
		●							
									●
		●							
									●
									●
		●							
									●
									●
		●							●
									●
									●
		●							
									●
									●
									●
									●
									●
									●
									●

INDEX

ABOUT THE AUTHOR

Doris Sloan is an Adjunct Professor in the Department of Earth and Planetary Science at UC Berkeley. For over 20 years she has taught Environmental Sciences for UC Berkeley and courses on the geology of California and the Bay Area for University Extension. She has led field seminars for the Yosemite Association, the Point Reyes National Seashore Association, the Sierra Club, and other organizations. Her current research focuses primarily on the sediments beneath San Francisco Bay and what they can tell us about the Bay's geologic history.

ABOUT THE PHOTOGRAPHER

John Karachewski has conducted geology projects throughout the western United States. He works for an environmental consulting company and also teaches adult education and community-college classes. He is an avid hiker and enjoys photographing landscapes during the magic light of sunrise and sunset. Examples of his images can be viewed at www.geoscapes photography.com.

Series Design:	Barbara Jellow
Design Enhancements:	Beth Hansen
Design Development:	Jane Tenenbaum
Cartographer:	Eureka Cartography
Composition:	Jane Tenenbaum
Indexer:	Thérèse Shere
Text:	9/10.5 Minion
Display:	Franklin Gothic Book and Demi
Printer and binder:	Everbest Printing Company

Introduction to California Desert Wildflowers, Revised Edition, by Philip A. Munz, edited by Diane L. Renshaw and Phyllis M. Faber

Introduction to California Plant Life, Revised Edition, by Robert Ornduff, Phyllis M. Faber, and Todd Keeler-Wolf

Introduction to California Chaparral, by Ronald D. Quinn and Sterling C. Keeley, with line drawings by Marianne Wallace

Introduction to the Plant Life of Southern California: Coast to Foothills, by Philip W. Rundel and Robert Gustafson

Introduction to Horned Lizards of North America, by Wade C. Sherbrooke

Introduction to the California Condor, by Noel F. R. Snyder and Helen A. Snyder

Regional Guides

Sierra Nevada Natural History, Revised Edition, by Tracy I. Storer, Robert L. Usinger, and David Lukas